T0219941

From Aardvarks to Zooxanthellae

John C. Avise

From Aardvarks
to Zooxanthellae

The Definitive Lyrical Guide
to Nature's Ways

 Springer

John C. Avise
Department of Ecology & Evolutionary Biology
University of California
Irvine, CA, USA

ISBN 978-3-319-71624-4   ISBN 978-3-319-71625-1   (eBook)
https://doi.org/10.1007/978-3-319-71625-1

Library of Congress Control Number: 2017960843

This Springer imprint is published by Springer Nature
The registered company is Springer International Publishing AG
The registered company address is: Gewerbestrasse 11, 6330 Cham, Switzerland

ENCOURAGING WORDS
To Joan, my better half,
Who relishes my daff.
Many nights in our bed,
She's snidely said:
"Very funny. You make me laugh."

Contents

Table of Contents (Extended)

ICHTHYOLOGY

Everything you wish,
From sea to dinner dish;
   How they have scales,
   Totally unlike whales;
We'll learn a lot about fish!

THE LIFE OF A FISH

Here we'll closely examine,
From the Eel to the Salmon,
   From freshwater to sea,
   Even "male pregnancy";
We'll try to keep you laughin'!

ENTOMOLOGY

Here we will inspect…
Many an "insect";
   Six legs on each…
   Is what we'll teach,
And much more we'll dissect!

INVERTEBRATE BIOLOGY

We'll examine each invertebrate,
And consider how they rate,
   From Clams to Octopi,
   And habitats they occupy,
Every such animal we'll celebrate!

MICROBIOLOGY

Every teeny tiny beast,
Can be cool, to say the least;
   From plankton in the sea,
   To symbiotic Microalgae
All the way to Mold and Yeast!

BOTANY

Plants here are mostly neglected,
'Cause the last time I inspected…
   They aren't even animal,
   And, hence, are quite dull;
That's why they've been rejected.

ANATOMY, PHYSIOLOGY, AND MEDICINE
These topics are self-explanatory,
But we'll explain them anyway;
   At the end of the day,
   You'll be able to say,
You understand them thoroughly!

ECOLOGY
From land to air to sea,
We'll cover exhaustively;
   And we'll even range
   Into climate change;
During the 21st century!

ETHOLOGY
Here we'll often rave…
On how animals behave;
   We'll even make sense…
   Of bioluminencense;
Everything you might crave!

EVOLUTION
It's hardly a delusion…,
This concept of evolution;
   For many of life's questions
   It offers helpful suggestions,
Or even complete resolution!

GENETICS
We'll cover inheritance
From many a stance
   Want to know about mitosis,
   Chromosomes, and meiosis?
Well, here's your chance!

ANTHROPOLOGY
Where did we come from?
Does this query seem dumb?
   Of course, my dear,
   God put us here;
But much deeper we will plumb!

APPENDIX

This silly little appendix…
Is included just for kicks;
   Many an animal phrase
   Warrants at least a gaze;
In our memory it sticks!

The Book

Hundreds of animal species provide the cast of characters for these newly composed bio-limericks, arranged into 17 chapters by taxonomic group (such as Birds, Fishes, Insects) or biological subject (such as Ecology, Genetics, and Anthropology). Sometimes multiple verses on one organism or topic provide an extended story-line across successive poems. In addition, several stylistic vignettes recur throughout the book, such as: (a) "On the Farm", which ranges from barnyards to fish farms to oyster farms; and (b) "Let's Play Jeopardy", where the reader guesses an animal from poetic clues the author provides.

Each little jingle can be read as a stand-alone offering a quick chuckle or biological insight. But watch out—these poetic tidbits can be as addictive as popcorn, such that some readers will feel compelled to consume each chapter and indeed the entire book at one sitting! Covering nearly every creature that any amateur or professional biologist has ever heard of, these pun-filled limericks provide humorous insight into each critter or its peculiar habits, in a sharply witty and cutely informative way.

The Author

Normally a far more serious scientist, John C. Avise is a Distinguished Professor of Ecology and Evolution at the University of California at Irvine. A member of the National Academy of Sciences, The American Academy of Arts and Sciences, and the American Philosophical Society, Avise has published more than 350 scientific articles plus 30 books on a wide range of biological topics related to genetics, ecology, evolution, natural history, and the human condition.

Mammalogy

AN ELEPHANT'S MEMORY
His head, with impressive weight,
Leads people to overrate...
    The Elephant's brain,
    Where memories reign,
But futures are left to fate!

FARSIGHTED
Which animal in a zoo
Can claim the highest I.Q.?
    Just query the staff—
    "It's the Giraffe,
At least from his point of view!"

ELKHORN

The Elk, with his thunderous bellow,
To his cows was an impressive fellow;
   But their awe turned to scorn,
   When he lost his bullhorn,
And his bugle came out like a piccolo!

BUTCHER'S MISTAKE

A Bull reared for meat has a fate…
Different from his milky mate.
   To a Cow's dismay,
   A butcher one day,
Executed a fatal miss-steak!

SYSTEMATIC DILEMMA

It's taxonomic drama
Distinguishing the Llama,
   From Alpacas who…
   Also live in Peru,
Thereby causing a di-llamma!

ILLEGAL IMMIGRANTS

The Mexican Armadillo
Was a sneaky picadillo,
   Who emigrated
   'Cross borders ungated
All the way to Amarillo.

CAMOUFLAGE

Deserts have many mirages,
Such as Camel entourages.
   Their humps 'bove the plain,
   Look just like a chain…
Of mountains-- perfect Camelflages!

DEN MOTHER

When days of winter are numbered,
A Bear gives birth, fully slumbered.
   The Cubs she's beared
   Must be fully prepared…
For mom to sleep on, unencumbered!

HYSTERIA

An uproarious dog, the Hyena,
Laughing across the savannah;
  Side-splitting guffaws,
  But it does give one pause:
Will he get a Hyena hernia?

ASSES

One bad thing about Jackasses:
They're stubborn and slow as molasses.
  When you want them to go,
  They often say "no";
You might as well leave them just as is!

MOUNTAIN GOATS

They climb cliffs precipitous,
In games highly treacherous.
  But, somehow, they know,
  That with sleet and snow,
It's time to get less participous.

HIDE-SEEKING

A Zebra is truly a Horse
Of another color, of course.
  With stripes black and white,
  She hides day or night:
By this cryptic tour-de-force!

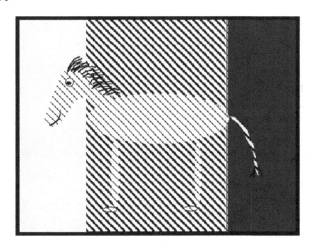

CENSUS OF SENSES

Hearing and vision are senses…
Used in mammalian defenses.
    Deer listen mostly,
    Bear watch closely,
But Skunks walk right up and scent us!

OXEN MORON

"Dumb Ox": a phrase used in fun…
For these cattle that weigh a ton.
    Thus, ironically,
    A "smart Ox" would be:
An intelligent Oxymoron.

WHAT THE HECK IS IT?

This mammal created a fuss,
Even driving people to cuss.
    It lays eggs, and has bill like a Duck;
    Taxonomists said: "What the ____? (heck)".
'Twas the Duck-billed Platypus!

THE BLUE WHALE

When the largest creature on Earth…
Gets pregnant and must give birth,
    It can't be much fun:
    The Calf weighs a ton,
And is more than 10 feet in girth!

A PRICKLY PAIR

Both a boy and a girl Porcupine
Are covered with many a spine;
    They like to date,
    But seldom mate.
So, for love they endlessly pine!

GROUP NAMES

Porcupine groups are "prickles";
Which gives me humorous tickles:
    What would two prickles be?
    Well, ironically:
Hundreds and hundreds of prickles!

THE LONG-NOSED ECHIDNA
Would I kid 'ya…
About the Echidna?
Its tongue and beak…
At their capacity peak,
Of ants, will happily rid 'ya.

LET'S PLAY JEOPARDY
His TV surname is Cleaver;
The pitch he works at is fever;
He builds lodges and dams,
And has thousands of fans.
Question: What is the Beaver?"

CAT FAMILY
A Bobcat's not tiny, yet shrinks
Compared to a Tiger, methinks.
A Cat seldom seen
Evolved in-between:
The proverbial "missing Lynx".

MISSING LYNX

RIVER OTTERS AND SEA OTTERS
Two very close species of Otter
Prefer either fresh or salt water.
But when seen in a zoo,
Which one? Here's a clue:
It's either one or the otther!

PRIDE AND GROOM

The Lion patriarch died,
So, his widows advertised.
  A young male responded,
  And soon they were bonded.
They found their groom, and he his pride!

WORDY WORDS

Some redundant phrases reiterate.
Take the Lion, a potent potentate;
  When his prey fell prey,
  We stayed way away…
While the consummate consumer ate.

RATHER HORNY

Behold the horn of the Rhino;
Hairs matted much like twine. Oh,
  What sort of a maniac
  Uses this aphrodisiac?
Don't look at me; How should I know?

BE FAST OR FAST

The macho feline philosophy
Entails a lot of ferocity.
  But Cheetahs well know,
  It requires also:
A healthy dose of velocity!

JUMPING OFF A CLIFF

For over-crowdedness stemming,
You could hardly out-do the Lemming.
  Proverbially,
  They jump into the sea…
Without any hawing or hemming!

OOPS!

The Dolphin could not have felt worse,
After bumping the boat off its course.
  His splashing around
  Made the ship run aground;
But it really wasn't on Porpoise!

GROUP NAMES

Dolphins swim in a "Pod",
Which is really rather odd;
  Though tightly packed,
  They never are stacked...
Like so many peas in a pod!

THE HARVEST MOUSE

Reithrodontomys raviventris:
His name is longer than he is
  Fortunate we are
  That he's not a star;
It would overfill the marqueses.

THE UGLY AARDVARK

Pig-like snout, floppy ears,
And rat-like tail in arrears;
  How the homely Aardvark
  Got aboard Noah's ark...
Was not by good looks, it appears!

JUST PLAIN LOVE

They crossed the plains on a lope,
The couple bursting with hope.
  They were soon out of sight,
  Chasing dreams in the night.
That's how Antelopes elope.

THE AMERICAN PRONGHORN

A couple of Pronghorns
In total have four horns.
  But ample speed
  Meets their every need...
By leaving each predator forlorn!

JUDICIOUS SETTLEMENT

For Roos seeking a divorce,
An option of last recourse...
  Is to have their property,
  Divided up properly...
By writ of a Kangaroo court!

SEAL WITH A KISS

Seals must be coy on their dates,
To carefully choose correct mates;
  Before "seal of approval",
  Comes thoughtful perusal;
Sea Lions would be a mistake!

WHITE POLAR BEARS

Their outlook got highly polarized
By the dreary place they colonized.
  No forest or grassy knoll
  Is near the North Pole;
So, these Bears became non-colorized.

PRAIRIE DOGMA

The motto in Prairie Dog town?:
Never face life with a frown.
  Approach every day
  With work and play.
Keep spirits up, burrows down!

NEWSPAPERS: FOOD FOR THOUGHT

My Dog thinks he's paying his dues,
When my paper he fetches and chews.
  His jaws are in spasm
  From enthusiasm;
He's going to digest the news!

MELANCHOLY BABY

The Fawn was no longer jolly;
The melon he ate proved a folly.
   There just wasn't room in...
   His overstuffed rumen.
So, he got watermelon-choly!

THE ART OF CHOKING

There once was a hungry Goat
Who got food stuck in his throat.
   With the veggie he vied,
   'Till he gagged and died--
Couldn't master the artichoke!

LOT OF BULL

He's a suitor most capabull,
Promiscuously sociabull.
   Whenever he's stuck
   In a mating rut,
He's a lover insatiabull!

MILK CARTONS

Selection got a bit zealous,
But of Cows we needn't be jealous;
   With bags hanging down,
   And long teats all around,
They look udderly ridiculous!

SPOTS BEFORE OUR EYES

In the dappled forest light,
This predator blends out of sight.
   ...Any Ocelot
   Scares us a lot
When he's startled, we get the fright!

A BUFFALO HELLO

The Bison was quite a load...
For the Puma to grab and hold.
   With the Cat gored and stomped on,
   The Bison said "Bye son,
You've just been Buffaloed."

GROUP NAMES

A Bison group is a "herd"
As of course, you've heard;
    Before White Men came,
    They could cover a plain…
In numbers that now seem absurd!

A BOAR AS A BORE IS ABHORED

Wild Hogs we mostly abhor.
Even sows find them hard to adore.
    They're homely and hairy,
    Their razor tusks scary,
And their piggish manners a bore.

GROUP NAMES

The words are meant to dazzle,
But generate unneeded hassle:
    'Sounder', 'drift', and 'drove',
    Are a wordy treasure trove
For a group of Hogs known as "passel".

A KICK OUT OF LIFE

Their hoof is a toenail like lead;
One kick and a Lion is dead.
    When pursued on the plain,
    Gnus have an aim
To land a nail right on the head!

SHOWDOWN AT THE CORRAL

'Twas a head-butting duel of note:
To the victor—two Nannys, one gloat.
    A Ram gained morale
    At the OK corral;
When Billy the Kid got his Goat!

DOG SIGNS

They do their business, alas,
Whenever my yard they pass;
    So, heed all these letters,
    You pointers and setters:
"DON'T LITTER! (and keep off my grass)".

GROUP NAMES

Wolf puppies comprise a "litter",
Whose poopies their den can litter.
  How do parents manage…
  To purge this baggage?
I'm asking this question quite litterally!

GROUP NAMES

Adult wolves hunt as a "pack",
The better a Moose to attack;
  Or maybe a Bear,
  Or a herd of Hare;
Odds in their favor they stack!

RICHLY DECERVID

Deer are entrepreneurs, you know;
Go-getters who reap what they sow.
  She chases the Buck,
  And, with any luck,
He ends up with plenty of Doe!

VISIBLE ENDITY

Those modeling it are 'a-plenty,
This visible hind entity.
  Why do Rabbits and Deer
  Have white on their rear?
Because hindsight is 20-20!

RABBITS HABITS

With reproductive expedience,
They litter at every convenience.
   With kids all around,
   Bunnies have found…
They love a Hare-raising experience!

GROUP NAMES

A group of Rabbits is a "warren",
But I should give you a warnin':
   It's also a "colony",
   So, I wonder if the…
First name above is warrented!

WOMANATEE

The Sea Cow or Manatee
Jealously guards her vanity;
   To a Manatee boy,
   She's a curvaceous joy;
It's an attribute of her womanity!

INSTINCT

A Skunk needn't stop to think;
As fast as the eye can blink,
   He raises his tail
   When under travail,
And relies on his sense of in-stinkt.

THE DAZZLING GAZELLE

A graceful Gazelle
Can run like hell.
From a Lion;
I ain't lyin.
So, all ends well!

A WILD WILDEBEEST

What's new?
The Gnu.
A Blue Wildebeest;
What an odd beast!
Who knew?

BATS WING THEIR WAY

They love to hang around,
Especially underground;
From caves they take flight
Into the night,
Guided by ultrasound.

GROUP NAMES

Groups of Bats are a "cauldron",
And, indeed, are like a cauldron…
Boiling with hot action,
And lots of compaction…
When exiting caves by the squadron!

AERIAL DISPLAY

Lethargically rests the Bat,
When hanging from this or that;
But when nighttime comes,
She quickly becomes
An aerial acro-bat.

SHE IS WHAT SHE EATS

Her name couldn't be neater,
Nor shorter, nor sweeter.
She is what she does
Simply because
She's an Anteater!

WEASELS AND FERRETS

Whatever troubles they're in,
These crafty cousin-like kin,
    Either Weasel their way,
    Or Ferret their way,
Respectively, out of 'em.

A BULL MOOSE

Stay away from any Moose,
Who happens to be on the loose.
    He has awesome strength,
    And a great deal of length
From his muzzle to his caboose!

YES, SHE HAS HORNS TOO

When I first saw a Longhorn Cow,
I couldn't help but say "Wow!"
    She had horns five feet wide,
    Which made me decide
To exit that pasture right now!

ANOTHER LONGHORN

And as for a Longhorn Steer,
He likewise elicits fear.
    He's so frustrated
    From being castrated,
That you had better steer clear!

NO, IT'S NOT SASQUATCH

The Orangutan is very…
Big, and strong, and scary.
    Oh, I almost forgot,
    He's also got…
A coat that's orange and hairy!

A GORILLA IS A REAL GORILLA

For this ape, the common name…
And the Latin name are the same.
    The awesome Gorilla:
    Gorilla gorilla.
Such redundancy seems inane.

MARSUPIALS VERSUS PLACENTALS

Said the parents of a young Wallaby,
"You can be whatever you wanna be."
  But the Joey knew
  That this wasn't true:
A placental he could never be!

MAMMAL OR SHARK?

The Orca is not a shark;
The difference is rather stark:
  He's much too debonnaire,
  And breathes lots of air,
In daylight and after dark.

SEA LION OR SEAL?

In truth, why would I be lyin',
'Bout Seal vis-a-vis Sea Lion?
  A Sea Lion hears
  Through external ears
So, a Seal is harder of hearin'!

APE OR MONKEY?

This Primate is thin as a ribbon;
Almost anorexic (I'm fibbin').
  Though a Spider Monkey
  Is both slim and funky,
So too is an Ape called the Gibbon.

GROUP NAMES

Most Monkeys live in a "troop",
Which is a name for their group;
  You could call it a gang,
  But that would be slang;
To such slander, you should not stoop!

A TOMCAT

He's hung out in thousands of dives,
Mated with hundreds of wives,
  That's not too bad,
  But remember he's had…
In total exactly nine lives!

THE POCKET GOPHER

The homely Pocket Gopher
Burrows like no other;
   He missed the memo,
   Or else he would know:
"When you're in a hole, stop digging"!

A NAUGHTY BADGER

Don't be mad at her,
Castigate or adage her.
   Without a doubt,
   She's no Boy Scout;
Thus, can't be a merit Badger!

GROUP NAMES

A Badger group is a "cete",
And I say without deceit,
   That to get less of them
   Or their numbers to stem,
You can't simply push "delete"!

THE PECCARY

The wife of a Peccary
Can be mighty scary.
   Especially to him,
   When vigor and vim…
Make her too hen-peckery!

THE RACCOON

Our nocturnal Raccoon
Can't be gone too soon;
   Harvesting what he can,
   Noisily from a trash bin.
Under a Harvest moon!

SLY LIKE A FOX

The Sunday was sunny,
And the Fox felt funny.
   It was not his habit
   To set free a Rabbit.
But this was the Easter Bunny!

IN A FAMOUS FOSSIL BED
He was a Horse, of course, of course,
But far from able to stay the course.
   Eohippus was a fossil:
   All bones and no muscle.
He could never beat a Racehorse!

CATASTROPHE
A Lion hunter quite oafy
Luckily bagged a trophy.
   But he got really miffed
   When the taxidermist
Created a cat-ass trophy!

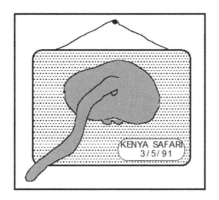

STATE TROUPERS
Many troups of Baboons
Act like groups of goons:
   Pillaging, scavenging,
   And generally ravaging
Kenya's prairies and dunes.

SAFARIS
While on the topic of Kenya,
What else can I tell 'ya?
   Zebras and Wildebeest
   And many another beast.
Aren't they totally cool? Booyah!

AUSTRALIAN MARSUPIALS

Do you ever wonder
What resides Down Under?
   It's really true:
   Many Kangaroo,
And Wallabies in great number.

LET'S PLAY JEOPARDY

He's covered with white hair
Huge shoes he likes to wear,
   He must outfox
   The Arctic Fox.
Question: What is Snowshoe Hare?

PLAYING GOD

God must have had a ball
Constructing the Narwhal.
   Half Unicorn, half Porpoise,
   Presumably on purpose,
Taking expertise plus gall!

BIGFOOT

Saskuatch, Yeti, Yeren, Yowie;
It makes you wonder just how he…
   Got so many labels,
   With no one yet able…
To prove he exists, actually!

LET'S PLAY JEOPARDY

They live in Madagascar,
At least 80 species for sure;
   Some have rings on their tail,
   Both female and male.
Question: What is the Lemur?

ON THE FARM

To fall into a deep,
And restful sleep.
   A home remedy
   That's totally free
Entails counting Sheep.

ON THE FARM

Though never done before,
This should work, I'm sure:
   To tally your herd of Cow,
   Here's approximately how:
Count teats and divide by four!

LET'S PLAY JEOPARDY

They're vastly bigger than us,
Both sexes have ivory tusks;
   They live in high latitudes,
   And have feisty attitudes.
Question: What is the Walrus?

THE SHREW

Is the Shrew
Very shrewd?
   Or naïve?
   I believe...
It's the two.

THE MOLE

When a Mole
Digs a hole,
   It becomes his house
   Like that of a mouse
On the whole.

THE MARMOT

I almost totally forgot
The Marmot on his rock;
   Short legs, bushy tail,
   A fat terrestrial Squirrel:
A Groundhog or Woodchuck.

TREE SQUIRREL

I'm often in the mood
To watch this little dude,
   As he scurries around,
   Planting acorns in ground;
Squirreling away his food.

ON THE FARM

Don't ever get downwind,
Or you'll really catch wind…
    Of Pig manure,
    That's for sure,
Emanating from his hind-end!

THE CHIPMUNK

Whenever your spirits have sunk,
Just stop and watch a Chipmunk.
    She's so cute and adorable
    You'll soon be quite able…
To lift yourself out of your funk!

THE BEACH MOUSE

The Floridian beach Mouse,
In sand-dunes digs his house,
    On waterfront property,
    Situated quite properly:
Ocean views but nary a douse!

ON THE FARM

Having many Cats
Means fewer Rats,
    And Shrews,
    And Mice too.
But not necessarily Bats!

THE VOLE

The Meadow Vole
Resembles a Mole,
    With small paws
    Rather than claws.
Both live in a hole.

ONE BIG CAT

Mountain Lion and Cougar,
Catamount, Panther, Puma;
    All different cognomens
    For identical specimens
With the Latin name *Puma*.

LET'S PLAY JEOPARDY

It's a top-end car;
It can travel afar.
　　In dim rain-forest light,
　　It's a hair-raising sight.
Question: What is a Jaguar?

THE COYOTE

Some call it a "kai-oat."
Others say "kai-oh-tee".
　　In any event,
　　It's evident (or evidently)
The size of a Goat (or Goaty).

THE FOX

He's fine at catching a Mouse,
But for other jobs he's a louse;
　　I note in particular,
　　He's not up to par,
For guarding a Hen-house!

THE COATIMUNDI

Furry Coatimundis,
Are also called Coatis.
　　If their name shortened more,
　　They'd become outerwear:
A "Coat" over their undies.

THE MINK

What do you think
About the Mink?
　　We all concur,
　　He's got great fur.
But might he be a fink?

THE WOLVERINE

The Wolverine
Is seldom seen
　　Very secretive,
　　Where it lives.
(So it would seem).

FLABBERGASTED

The Elephant Seal
Has little appeal.
    From his long flabby nose,
    To his flipper-like toes;
So much fat it's almost surreal!

CARIBOU

I'd really like to hear
Whether Caribou or Reindeer…
    Pull Santa's sleigh
    On Christmas day.
Just before each New Year.

THE MUSKOX

A feisty herd of Muskox
Will really knock off your socks.
    When a Calf is in danger,
    They form a safe manger…
By encircling it like a box!

LET'S PLAY JEOPARDY

It was a Chevy car;
It lives in Africa;
    Don't be a dope,
    It's an Antelope.
Question: What is an Impala?

LET'S PLAY JEOPARDY

Its name rhymes with "slang"
It often herds in gangs
    It's a Ford product
    With sales hard to top.
Question: What is a Mustang?

MULES

Even after these hybrids go feral,
They're still caught under a barrel;
    Whether captive or loose,
    They just can't reproduce.
That's their fate, being sterile!

AN AMERICAN MARSUPIAL

The American Opossum
Is totally awesome,
 Until he's dead,
 Or perhaps instead,
He's only playing possum!

THE BORDER COLLIE

The Collie was confused:
Was he just being used?
 Was he master of the herd,
 Or a slave of the shepherd?
In truth, both roles were fused!

BEING SHEEPISH

This common situation…
Occurs across our nation:
 Ram and Ewe
 Sit two by two
In a flock (or congregation).

COWS' GRAZING HABITS

The field came highly lauded:
Its fresh grass was applauded.
 Did they change their postures?
 Maybe seek greener pastures?
No-one goes anymore: it's too crowded!

Ornithology

UNGAINLY

The long legs and neck of a Stork…
Provide much bodily torque…
 For mating display,
 And capture of prey;
But the guy sure looks like a dork!

WHO?

Owls are said to be wise.
Not so; it's all a disguise!
 I've looked and it's true:
 They're not in *WHO'S WHO*--
So, someone's been spreading some lies!

© Springer International Publishing AG 2017
J.C. Avise, *From Aardvarks to Zooxanthellae*,
https://doi.org/10.1007/978-3-319-71625-1_2

GROUP NAMES

An Owl flock is a "parliament";
I never knew what that meant.
   So "parliamentary procedure"
   I suppose must refer…
To how hoot-Owls rage and vent!

ARCTIC TERN-OVER

They go to considerable bother,
In travels unlike any other.
   From North Pole to South,
   Their migration is about;
One good Tern begetting another.

WHOOPING CRANES AND SANDHILL CRANES

The Whoopers are nearly extinct.
Their fate to the Sandhill's is linked.
   Via the latter's foster care,
   On a hope, wing, and prayer,
Whoopers might be less on the brink.

ELECTRICITY IN THE AIR

The Swallow-tailed Kite
Is an awe-inspiring sight:
   As he soars overhead…
   Through a thunderhead,
In electrifying flight!

PROPER NAME

The Bahama Bananaquit…
Has a name that's appropriate.
   He lives in the Bahamas
   And refuses to eat bananas.
But why not then Brusselsproutsquit?

PEREGRINE FALCON

With sublime acceleration,
This raptor went on vacation…
   From mountain to sea;
   'Twas a quick journey:
A short Falcon peregrination!

AWKWARD

The Great Auk, a magnificent bird,
Went extinct, as I'm sure you've heard.
  The main reason why?--
  He just couldn't fly.
Quite a penance for being Aukward!

BYE BYE, DODO

An insulting duplicitous word,
Double-jeopardized this stupid bird.
  Its I.Q. was low,
  So, even more so...
Its maltreatment should leave us disturbed!

LET 'EM FLY

I hate to be unduly derogatory,
But birds are indiscriminatory;
  'Round my car they fly low,
  Releasing globs of guano;
They must think it's a lavatory!

AVIAN RESTROOMS

When a Gull must go in a hurry,
It seldom needs to much worry.
  The loo it searches for,
  States right on the door:
Either Gulls or Buoys.

UNDERTAKING UNDERTAKING

Pending evidence to the contrary,
These birds think their prey is wary.
  It lies quietly on the ground,
  Without movement or sound:
It's a carcass, the Vulture's quarry!

SWELL SWILL

To us, rotting meat tastes like hell,
But Vultures think it's really swell.
  So, carry on,
  Eating carrion,
And relishing each morsel!

EAGLES AND ILLEGALS

Erns were ill, causing men to enact…
Bans on DDT; That's a fact.
    I guess no lawyer saw,
    That the law broke the law…
By enacting an ill-eagle act!

GROUP NAMES

Eagles gather in a "convocation",
Which is more than an avocation;
    Feeding on rivers and lakes,
    With all the effort this takes,
Qualifies as a full vocation!

A MATTER OF OPINION

In the avian plumage dominion,
Of feathers, birds have 'bout a million.
    But whether they're grounded
    Can strangely be founded…
On merely a matter of a pinion!

ONE TOUCAN, TWO CANNOT

A Toucan tried to supplant
Another perched on a plant.
    But the branch broke in two,
    Causing both birds to rue:
What one toucan do, two can't!

HOLDING IT IN

An Ostrich, without any doubt,
A melody someday will spout.
    I cannot be wrong
    That he's bursting with song--
'Cause little has ever come out.

CUTE LITTLE CUSSES

With their sweet serenades 'a blaring,
Tiny male House Wrens are so caring.
    That's the naive view;
    Here's what's really true:
They're mostly cussing and swearing!

PENGUINS

Since Penguins can't speak, I'll articulate
A falsehood they'd like to eradicate.
  Penguins never are found...
  Northern Hemisphere round.
Their range is strictly Antarcticate!

GROUP NAMES

Many seabirds inhabit a "rookery",
Which is a poor place for a rookie;
  Adults fight and mate,
  Then impatiently wait...
To lay eggs in each cranny and nookery!

HEALTHY AND SHINY

A Drake need not contemplate whether,
To spread its preening oil on a feather.
  Be it sun, sleet, or rain,
  He remains dry and vain...
Feeling Ducky whatever the weather!

GROUP NAMES

A group of shot Ducks is a "brace".
For which you could make a case...
  For a truly mass murder,
  By some zealous "birder".
Perhaps a stiff sentence he'll face!

GROUP NAMES

Many Ducks on the water form a "raft",
Which is a well-evolved feeding craft;
  Despite lots of diving,
  The raft stays thriving,
By scanning for predators, fore and aft!

DAFFY, STUFFY DUCKS

Our feathered friends in the swamp,
Where the weather is very damp;
  In processions of waddles...
  March 'round their puddles,
Full of circumstance and pomp!

GROUP NAMES

A bunch of tame Geese is a "gaggle"...
When they walk with their signature waggle.
   How many is a bunch?
   I have a good hunch:
At least four, but let's not haggle!

GROUP NAMES

In flight, these same Geese are a "skein",
Which is where they sometimes are seen.
   I guess a gaggle on the ground
   Was inadequate, hunters found.
So, what are Geese called if they're skiing?

GOONY BUFFONERY

Boobies are easy marks...
For slightly off-color remarks.
   They hang out in pairs,
   Displaying their wares,
And igniting pre-nuptial sparks!

LOONACY

A humorous bird is the Loon,
When he yodels his laugh-like tune.
   Yet how I long
   For his haunting song...
Under a northern moon.

THE RUDDY TURNSTONE

Whether instinctual or learned,
His name he has duly earned...
   As he probes pebble beaches,
   An important lesson he teaches:
Never leave a stone unturned!

THE SANDERLING

This darling little Sandpiper,
Runs in a constant scamper.
   He forages on sand...
   'Twixt the sea and the land,
Dodging waves and acting hyper.

SHOREBIRD FEEDING FITS THE BILL

Bills that bow up are recurved;
Those arching down are decurved.
   Some are long, short, or straight,
   But whatever their shape,
They all probe where dinner is served!

FLIGHT

When birds are in V-formation,
It takes little imagination,
   To see just why
   It helps them fly
Straight to their destination.

MORE ON V-FORMATION

When a group flies in a "V",
There's often an asymmetry.
   For this simple observation,
   There's a clear explanation:
Longer arm, more birds, simply!

DUAL LIFESTYLE

On her nest inside a tree hollow,
Quietly rests a Tree Swallow.
   But once in the air,
   She darts here and there,
With movements hard to follow.

WHITE-THROATED AND CHIMNEY SWIFTS

Every Swift
Is very swift:
   Fast in flight,
   Day or night.
It's quite a gift!

FISH-EATING HAWK

When pesticides enter in a waterway,
They lodge in the prey of an Osprey.
   So, it can't be assumed...
   That from fish he's consumed,
He won't get sick, let us pray!

THE MOCKINGBIRD

In my yard, I hear him presently;
He started out quite pleasantly.
    But it's early spring,
    And his urge to sing…
Makes him go on incessantly!

MORE ON THE MOCKINGBIRD

He might mimic another bird,
Or some other sound that he's heard.
    From a Spotted Towhee,
    To a whistling pot of tea.
Sometimes it gets quite absurd!

THE AVIARY WAS A FULL HOUSE

One pair, two pair, three of a kind,
Let's get straight what's on my mind.
    The enclosure was flush
    With many a Thrush.
And all the Finches you could find!

THE WHITE-EYED VIREO

"Quick, bring me a beer, chick".
Or "chick, buy me a beer, quick."
    That's the song they describe
    In each birder's field guide.
But it's really "chick-per-weeer-chick"!

BROWN AND WHITE PELICANS

Most Terns and Gulls have a ban,
On fish bigger than a frying pan.
    Their bills are too small,
    So, they send out a call:
What they can't catch, Peli-can.

A PETREL'S PETROL

The Petrel is a flying machine;
Light, yet strong and lean.
    He uses Krill
    His tank to fill;
No need for gasoline!

HERONS AND EGRETS

A Heron or an Egret?
I often tend to forget.
   Which is which?
   So, sometimes I switch…
Their names, which I then regret.

MANY SPECIES OF SPARROWS

Some birders don't like sparrowing,
But that view seems to be narrowing.
   Though these birds look alike,
   Many ornithologists like…
I.D.'s when they are most harrowing!

PUFFINS

Sometimes it's bills they're stuffin',
And often it's feathers they're fluffin';
   Whatever their task,
   There's no need to ask…
Why are they huffin' and puffin'?

GROUP NAMES

A small group of Quail is a "covey",
In a setting that's extremely lovely;
   Hen, chicks, and a cock,
   Preening and taking stock,
With everyone acting lovey-dovey!

THE SKYLARK

The answer this question poses,
Is truly right under our noses.
   How many Skylark
   Were on Moses' ark?
Zero! (It was Noah's ark, not Moses'.)

ON A SUMMER FIELD TRIP

"What does a Gnatcatcher eat?"
This question was dropped at my feet...
   By a student who
   Clearly hadn't a clue.
She'd been too long out in the heat!

WHO'S BURIED IN GRANT'S TOMB?

"What color is a Blackbird?"
Is another question I've heard.
  "What kind of a beak
  Has an Evening Grosbeak?"
Such queries can get quite absurd!

NUTHATCHES

Some questions in academia
May seem perhaps even nuttier:
  "What kind of a cache
  Does a nuthatch stash?
Almond, pecan, or Macadamia?"

MORE STUDENT REACTIONS

There's always some wisecracker,
Who mocks a Clark's Nutcracker,
  And the Sapsuckers too
  Who get stuck in tree goo,
Such tacky jokes just get tackier.

AVIAN NOMENCLATURE

Some birds have such a long name
I admit it could drive you insane:
  Phainopepla, Pyrrhuloxia,
  It takes a lot out 'a 'ya,
When these you try to explain!

THE HAWK

He can see to the horizon…
With his spectacular vision.
  But Mice on the forest floor,
  Are what he mainly looks for…
And always keeps his eyes on!

GROUP NAMES

A group of Hawks is a "kettle",
With which you should never meddle.
  They have talon-like claws,
  And abide by nature's laws,
Don't ever get caught in the middle!

ON THE FARM
The coop had quite a batch,
As did the garden patch;
   With so many about,
   It was hard to count...
Our eggs before they hatch.

ON THE FARM
Though she tried,
She barely flied...
   Across her pen.
   Why? For her yen...
To get to the other side!

FLIGHTLESS BIRDS
What is a Ratite?
Well, hold on tight:
   Ostrich, Rhea, Emu, Cassowary;
   They're all quite extraordinary:
What they cannot display is flight!

LET'S PLAY JEOPARDY
This person lives in N.Z.
As a bird it can barely see,
   He's a little guy
   Who cannot fly.
Question: What is a Kiwi?

A NEW ZEALAND MORTUARY
The giant Moa
Don't live no moa.
   Why do I think
   It went extinct?
I saw it in a Moa-tuary.

GROUP NAMES
Swans aggregate in a "bevy",
Afoot or a-swim near a levee;
   They're aggressive and large,
   And they like to take charge...
By playing the role of a heavy!

MY BUDGIE

My Parakeet
Is very sweet
  But bless his heart
  He's not too smart
He only says "Tweet"!

GROUP NAMES

In nature, Parrots form a "company",
To fully enjoy each other's company;
  But in a pet store
  I'm pretty sure:
They're just working for the company!

ON THE FARM

On a Turkey farm, Gobblers and Hens...
Spend most of their lives in their pens.
  But on special holidays,
  At the end of their days...
They go dining out with new friends!

THE COMMON CROW

I want to crow
About the Crow:
  He's totally cool;
  And no-one's fool.
Just so 'ya know!

AVIAN TOOL USE

The Caledonian Crow
Is even smarter, though.
  He employs sticks
  To impale insects.
"Utensil use" on the go!

GROUP NAMES

A group of Crows is a "horde":
Such an evocative word!
  When this mob is around,
  So much sight and sound...
Assures that you'll never get bored!

ON THE FARM
Does a Chicken have only one breast?
This question can be put to final rest:
    Yes, it's one, not two,
    Unless cut in two.
So, next time you'll pass this test!

FACING THE GUILLOTINE
Next time you have a meal…
Of Chicken, here's the deal:
    While a Hen lays her eggs,
    A Rooster frantically begs:
"Mr. axe-man, don't make me kneel!"

THE PTARMIGAN
What is a Ptarmigan?
What? Come again?
    It's like a Chicken…
    On alpine Lichen.
Oh, I thought you said 'tarmigan'.

NON-KISSING COUSINS
The Crow had acted too brazen,
Snitching the food he was cravin'.
    He stole from his cousin,
    Who then started cussin':
"Nevermore" quoth the Raven!

NO, IT'S NOT PIGMENTS!
Why is a Blue Jay blue?
And why is the sky that way too?
    Both are due in effect
    To the "Tyndall effect"
Which scatters light of that hue.

HOW JAYS GOT THEIR NAME
Hey, Mr. Jay
What d'ya say?
    "Jay… jay"
    Every day;
Oh, okay, okay!

GROUP NAMES

A flock of Jays is a "party",
Which is very noisy and hearty;
   The birds carouse and play
   Several hours every day;
They really know how to party!

LET'S PLAY JEOPARDY

It has a deep pink glow,
It can steal a zoo's show;
   It has an extremely tall stance;
   Sounds like a sexy Latin dance.
Question: What is the Flamingo?

A WRONG-TERN COLONY

These languid birds were prone
To mostly stay right near home,
   And smoke marijuana,
   Whenever they wanna',
Thus, leaving no Tern unstoned!

THE AMERICAN ROBIN

In my garden, he goes hoppin'
His body and tail a-bobbin',
   His goal is firm:
   Find every worm.
Let's call him "rockin Robin"!

WATERTHRUSHES AND SPOTTED SANDPIPERS

As they scurry along a lakeshore,
You always know what's in store:
   They bob up and down
   Like some rockin' clown
It must satisfy them to the core!

THE AMERICAN DIPPER

He feeds wholly underwater
In a habit unlike any other.
   Picking tiny shrimps,
   And insect nymphs,
One after another.

MORE ON THE AMERICAN DIPPER

The American Dipper…
Always looks dapper.
  In his grey diver's suit
  Which looks real cute.
All he needs now is a flipper!

OUR NATIONAL EMBLEM

An Eagle
Is regal
  [Until he eats a Rat,
  or perhaps your Cat,
Or your Beagle!]

SWAN SONG

One decade, and most birds retire,
To join a more heavenly choir.
  They soar up and away
  Having no price to pay,
They just cash in their frequent flyer!

The Life of a Couple of Birds

WHO'S BRIGHT AND WHO'S DULL
Males tend to be brighter…
Than females, who are lighter…
　In their plumage, that is;
　Neither is a mental whiz!
She just stays out of sight more.

BEAUTY AND LICE
A handsome gent seldom fails
To attract more astute females.
　A plumage that's pretty
　Attests that he's lice-free;
No lady accepts lousy males!

PLAIN JANE
But then, on the other hand,
Each bachelor searches the land…
　For the drabbest of mates,
　Whom he fully appreciates
Because "gentlemen prefer blands".

LURING THE SUCKER
Around a suitor, she's sure
To always be sweet and demure.
　When fishing for mates,
　Those are tempting baits;
Just watch, as she casts her allure!

© Springer International Publishing AG 2017
J.C. Avise, *From Aardvarks to Zooxanthellae*,
https://doi.org/10.1007/978-3-319-71625-1_3

SILENT PARTNER

She's vocally often quiet,
While his babbling is a riot.
    In humans, of course,
    It's just the reverse.
Where the woman... (Whoops!, I better cram it!).

CROONING AND SWOONING

If he sings a sweet enough tune,
It may lead to a honeymoon.
    They can be quickly wed,
    Just a nod of her head…
And he changes to henpecked from groom.

THE WEAVERBIRD

In this case, the male knows best,
How to weave a fantastical nest.
    Thereby getting a first date
    With a prospecting mate.
After which she'll do all the rest.

HOUSEWIFE AND BREADWINNER

While the wife holds sway at the nest,
And can't have so much as a guest;
    Hubby's guarding their ground,
    Incessantly flitting all around;
(Also flirting, the girls can attest!).

EMOTIONAL PICTURE OF INFIDELITY

All of life's a big stage, that's a fact;
So, when her mate behaved without tact,
    He had to exit stage right
    In an action-packed flight:
His wife caught them right in the act!

THE BIG BIGAMY GAME

Birds are said to be mostly monogamous,
But for both sexes that's too monotonous.
    He enjoys polygyny,
    She likes polyandry;
So, they compromise: both are promiscuous!

WHO RULES THE HEN-HOUSE?

"King-of-the-roost" is oft spoken,
But the phrase is just a mere token.
   The real boss of the roost…
   Is the Hen, not the Rooster.
Her laws are not to be broken!

THE AVIAN EGG

Hard-shelled eggs are "cleidoic";
They evolved in the Mesozoic.
   For a chick to break out,
   There's surely no doubt:
It takes efforts nearly heroic!

THE AVIAN NEST

They come in many a form:
There is no standard or norm.
   From a scrape in the ground,
   To a mound-builder's mound;
They'll become the nestling dorm.

MORE ON AVIAN NESTS

It's hard to say what works best,
When it comes to an avian nest.
   Cups, hollows, and domes,
   All make excellent homes;
But other styles too pass the test!

A CAVITY NESTER

The European Starling
Is a homely little darling.
   And he's also a pest
   Stealing many a nest.
Why is that so startling?

NESTLINGS, SPECIAL DELIVERY

Then one day, who should they see,
But a Stork flying down to their tree.
   Birds have chosen the option
   Of this form of adoption.
For Mom, it's an easy delivery!

THE CLUTCH

An avian brood, or "clutch"
Can vary in size very much.
    From only two,
    To quite a few:
As in Ducks, Geese and such.

LEAVE ROOM FOR THE NESTLINGS

Each egg is the size of a pea,
Yet it takes only two or three
    To nearly fill up
    Each tiny teacup
Of a Hummingbird nest in a tree.

INCUBATION

Sitting on eggs night and day,
Is far from their idea of play;
    But to keep eggs warm,
    And safe from a storm,
It's perhaps the only sure way!

MORE ON INCUBATION

Both parents can incubate,
Which really works out great;
    While she sits on the nest,
    He can feed and get rest;
And *vice versa*, needless to state!

PIP-PIP HOORAY

When a chick in the dark wants a peek,
He uses a special pip-tooth on his beak…
    To crack open the egg,
    Emerge, peep, and beg;
Even though he's a mere little pipsqueak!

AVIAN HATCHLINGS

Shortly after they're "born",
Hatchlings come in two forms:
    Either naked and hapless,
    Or downy and less helpless;
Altricial and precocial are the terms.

TENDING THE CROPS

It's up to both parents to feed
Hungry young mouths all they need.
So, both Mom and Pop
Quickly plant in each crop:
Lots of millet and other birdseed.

A SEAGULL'S CROP OF YUMMIES

The hatchling impatiently waited
For repast enough to be sated,
Until much to his delight
Came the smell and sight
Of fish chowder regurgitated.

BROOD PARASITISM

"Egg-dumping" is another name,
For this strange reproductive game.
Brood parasitism
Is an odd realism
In the avian domain.

EGG-DUMPING

Each "brood parasite" has a quest:
Lay eggs in another bird's nest.
It's a devious game,
Done without shame;
Foster parents do all the rest!

MORE ON BROOD PARASITISM

To us, it's immediately apparent,
When a Warbler is not the true parent
Of any hatchling who…
Is an egg-dumped Cuckoo;
So, the Warbler's efforts are errant!

THE BROWN-HEADED COWBIRD

Cowbirds are brood parasites,
Dead set on setting their sites…
On duping their host
Into working the most;
Unconcerned about wrongs or rights.

FOSTERING DISRESPECT
His foster folks got not much:
Not even a "thank you", as such.
   Only eggs on their faces...
   When through their graces:
A Cowbird came through in the clutch.

SOLO, SOLONG, BYE-BYE
In three weeks, the family's on edge;
The youngsters perched on a ledge.
   They jump into space...
   To find their own place...
In nature. They're off, fully fledged!

THE FLEDGLINGS
Even after they're out of the nest,
Their parents still find little rest;
   The fledglings need care,
   Seemingly everywhere:
Each can really be quite a pest!

SEPARATION
You'd think after all they've been through,
Pairs would stick together like glue.
   Birds don't get divorces,
   Just go separate courses,
'Till next spring, when vows they'll renew.

BIRDS-EYE OVERVIEW
That's a bird's life: its ebb and its flow.
Not too bad, as creatures' fates go.
   All their joy and strife;
   It's a lot like the life...
Of a certain Primate I know.

Herpetology

SNAKE-SMITTEN
Inside their dens of iniquity,
Lay creatures of great antiquity.
    Rattlesnakes' homes
    Send chills through my bones.
I wish they weren't quite so ubiquity!

PIT VIPERS
Most Snakes lay eggs; they're oviparous.
Others are ovoviviparous.
    Their young are born live
    After hatching inside
But who cares! Just tell me who's viperous!

MORE ON PIT VIPERS
Cottonmouth and Copperhead,
Fill us with plenty of dread...
    In the summer season,
    And with good reason.
One bite could leave us dead!

UNCONSTRICTED ON THE ARK
Snakes such as Python and Boa
Owe a great debt to Noah:
    Who said "What the heck!
    Slither on deck."
But I can't say I'm glad to know 'ya.

© Springer International Publishing AG 2017
J.C. Avise, *From Aardvarks to Zooxanthellae*,
https://doi.org/10.1007/978-3-319-71625-1_4

RACERS AND COACHWHIPS

People can make mistakes
When identifying these snakes.
They both look like whips,
And as my wife quips:
"Who cares for Heaven sakes!"

IN THE GARDEN OF EDEN

He accosted the naïve lass,
And wouldn't let her pass.
Then he offered an apple,
And she had to grapple
With this evil Snake in the grass.

OFF TO A BAD START

There had to be some mistake,
Quite a lot was at stake.
On their honeymoon,
The bride and groom
Uncovered a Garter Snake!

COW PATTY

A Toad grew suddenly flatter,
When a Cow happened to sit on her.
Said the Cow with chagrin:
"My, don't you look thin."
This was rather cruel praise from the flatterer!

TOP FROG

Behold the noble Bullfrog,
On his throne atop a big log.
Who needs the sweet kiss
Of a maiden or princess?
He's already king of the bog!

GROUP NAMES

A group of Frogs is an "army";
Which is justifiably alarmy...
When it crosses a street,
And is likely to meet...
Busses that could be harmy!

THE TADPOLE

He couldn't make a peep,
Nor get a good night's sleep.
   His current life-stage
   Didn't fit the adage:
"Look before you leap!"

TOO MUCH TO DRINK?

I don't think this is slander,
About the Salamander:
   He can't walk a straight line,
   Which I suppose is fine.
He'd rather just meander!

GROUP NAMES

A "maelstrom" is many salamanders…
In a group in a stream, as it meanders,
   Or a pond, marsh, or lake,
   Or whatever it might take…
To attract such a crowd of bystanders!

LET'S PLAY JEOPARDY

This aquatic guy is cute,
Vocally totally mute;
   His surname is Gingrich,
   But he's not very rich;
Question: What is the Newt?

CAMOFLAGED CHAMELEON

The ever-changing chameleon
May seem like a comedian;
   But he alters his color
   So that any predator…
Is less likely to make a meal a' him.

SHELL SHOCKED

Inside his home, full of doubt,
A turtle glances about.
   The fellow's so scared
   That even if dared,
From his shell he'd never come out.

KEMP'S RIDLEY RIDDLE

Ridley Turtles have a strong mindset:
They all nest on one beach; no hedge-bet.
  Much to mankind's dismay,
  They may vanish someday,
Having placed all their eggs in one basket!

LET'S PLAY JEOPARDY

Its old Spanish name is Galápago,
And it lives in the Galapagos;
  Its shell is thick,
  It's not very quick.
Question: What is the Tortoise?

ALLIGATOR ALLEGORY

Did you hear the sad story
'Bout a Gator's female quarry?
  She went for a ride,
  But ended inside.
It was horribly alli-gory!

A BAD FIRST DATE

She didn't like his style;
In fact, she found it vile:
  "I'll see you later,
  Alligator;
I'm a Crocodile!"

PRAISE THE LORD!

The South American Caiman
Is a rather smallish specimen.
  Made like a Gator
  By our Creator.
For some unknown reason.

SHROUDED IN SHREWDITY

Barely visible in the fog,
In a habitat full of sog,
  Lays a Gator in wait
  For his dinner date;
Still as a lump on a bog.

ON THE FARM

A well-run Alligator farm
Actually does no harm…
  To the environment,
  So, time's better spent
Not sounding undue alarm.

ECTOTHERMS

"Their blood runs cold"
We've all been told;
  But the longer they sun,
  The warmer they become;
So, I'm not totally sold!

THE GILA MONSTER

This highly venomous lizard
Can sometimes be a hazard;
  But when it's cold out,
  They're seldom about,
You can safely hike in a blizzard!

GECKOS

The cute little Gecko
Is always on the go;
  Often indoors,
  From ceilings to floors
His padded feet grip like Velcro.

LAND AND TREE IGUANAS

"Hey, miss land Iguana
Let me lay some good lovin' on 'ya!"
  "NO!: get away from me;
  Go climb up your tree."
"In other words, I don't wanna!"

THE "HORNED TOAD"

This story may be a bit corny,
Why this lizard is so ornery:
  Will his frustration explode
  'Cause he's not a toad?
Or 'Cause he's just so darn horny?

NO, HE DOESN'T
What do you think
About the Skink?
Not much,
As such.
(Does he stink?)

ON THE FARM
About one farm in a million
Is exclusively reptilian.
Alligator farms
Have few barns,
Because ponds are a Gator's dominion!

MORE ON GATOR FARMS
Why raise Alligators?
For tourist spectators.
Plus, their rawhide pelts
Make fine boots and belts…
For apparel speculators.

MORE ON THE TORTOISE
It was a very odd pair;
The race didn't seem fair.
But the slower one won
The ten-kilometer run…
'Twixt the Tortoise and the Hare!

FEMALE SPERM STORAGE
Some Turtle females have learned
How to store a mate's viable sperm…
For three years or more;
But whose keeping score?
The exact duration's not firm.

MORE ON FEMALE SPERM STORAGE
The sperm that females store
In their reproductive core…
Came from a favored mate;
So, they don't further date.
And it sure beats being a whore!

PTEROSAUR

The first aerial dinosaur
Could just barely fly, much less soar:
     Despite exertions,
     On long excursions,
He crashed and became ptero-sore.

REPTILIAN FLIGHT

Nothing could inhibit…
Their entrepreneurial spirit;
     When Pterosaurs took flight,
     They lost all fear of height,
And the sky became the limit!

DINOSOARER

A fossiliferous mix
Was *Archaeopteryx.*
     From Dinosaur,
     To birds galore,
Halfway in-be-twixt.

BIG IMPACT

Dinosaurs were surprisingly stoic
Toward the end of the Mesozoic.
     A comet's iridium
     Finally did 'em in.
No chance to be real heroic!

GONE BUT NOT FORGOTTEN

Remember the Brontosaurus?
And likewise the Allosaurus?
     These reptiles are gone,
     But their names live on
Buried in my Thesaurus.

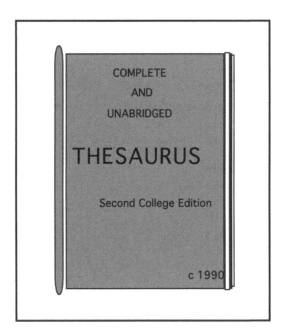

Ichthyology

FISHY SCIENCE

I've discovered the etymology
Of the field called "ichthyology".
　　From piscine disease
　　To their philosophies,
Or from "ick" to fish "theology"!

EVEN FISHIER

What could it possibly be?
The answer comes easily;
　　It has fins, scales, and tail;
　　No, it can't be a whale;
It sounds very fishy to me!

BODY AND SOLE

A very flat fish is the Flounder.
If he weren't so flat, he'd be rounder.
　　Well, he really is round,
　　Just built close to the ground.
He's a flat and round flounder arounder.

© Springer International Publishing AG 2017
J.C. Avise, *From Aardvarks to Zooxanthellae*,
https://doi.org/10.1007/978-3-319-71625-1_5

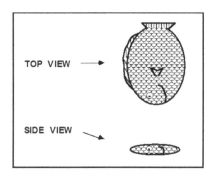

WHAT'S THAT YOU SAY?
Their eyes were made for peering,
Their fins designed for steering,
  And Fish hear just fine
  Through their lateral line;
Else, they'd be hard of Herring.

NET EFFECT OF FEELING INSANE
A Tuna feels anguish and pain.
He struggles for freedom in vain.
  He hasn't a guess
  Why he's in such a mesh.
Could it be that he's gone in-seine?

IN PRAISE OF RAYS
A Manta Ray glides here to there
With a grace and a beauty that's rare.
  His shape is flattened,
  So, clearly it's happened,
That flattery can get you somewhere.

GREAT WHITE SHARK
A shark lurks under the dock.
Of people, he takes careful stock.
  Any diver would make
  A fatal mistake:
He'd be in for a great white shock!

ELECTRIC EELS
This nightmare is really quite jolting:
All around, slimy fish are bolting.
  They charge up my dock,
  Giving shock after shock;
Electric Fish are revolting!

POP

A Pufferfish couldn't keep
From having this dream in his sleep.
  He'd huff and he'd puff
  'Till he blew himself up,
And got "rupture of the deep."

GAME EXCUSE FOR DANGLING LINES

To wardens, he stammered "But, but...
It was nothing deliberate."
  For no good reason,
  He'd angled off-season--
Just for the Halibut.

CRAPPIE DAY

There, floating by, just above her,
Was a succulent Minnow, in hover!
  She chomped down on the fish,
  But did not get her wish:
She could not tell the hook from its cover.

DENSELY PACKED

Two virile young Sardines at sea
Feel an urge for some intimacy.
  A big school they find,
  And achieve peace of mind;
Sardines simply abhor privacy!

GROUP NAMES

Any group of fish is a "school"
For which the Golden Rule...
  Is to stay densely packed,
  So that odds are stacked,
Against any predatory fool!

GROUP NAMES

A school is also a "shoal",
And has the identical goal:
  Stay tight together
  Even whenever...
A predator takes its toll!

VERILY HUNGRY

An Anglerfish, highly voracious,
Is a dinner host most ungracious.
    His guests don't discern
    That his lure's no worm,
'Till they fathom he isn't veracious!

ODIFEROUS ODE

An ode to the lovely Smelt--
Silvery, scaly, and svelte.
    But when dead in droves
    Along shores and coves,
Better they shouldn't be smell't.

POINTLESS

Swordfish have a point *a priori*;
So do Swordtails, *a posteriori*;
    The Brook Stickleback
    Has five points on his back,
But there seems no point to this story!

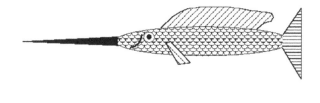

ZEBRAFISH: A FIGMENT OF OUR IMAGINATION?

Their striping is highly contingent
On how they distribute their pigment.
    That's honestly true,
    And, therefore, not due
To imaginary figment.

MORE PALATABLE NAMES
The Dolphin became Mahi-Mahi,
And the Tuna became Ahi-Ahi.
   These names are new
   On the chef's menu.
But they still make excellent sushi!

A FISHERMAN'S FISH
The Northern Pike
Is no little tyke.
   He's long and slender,
   And his meat is tender.
What's not to like?

THE ALLIGATOR GAR
This very large species of gar,
Is nearly as big as a car.
   Well, that's not quite true,
   But here's a clue:
He's bigger than scooters by far.

A SUCKER'S BIG MOUTH
This Fish latches on with it,
Sucking your eyeballs within it.
   If you think this is true,
   You haven't a clue.
Yes, there's one born every minute!

THE KOI
The Koi is an ornamental Carp,
Who was bred to look really sharp:
   Yellows, reds, and blues,
   And other bright hues.
If I found any faults, they'd be carps!

CHUBBY LITTLE CHUB
If you like the Chub,
Join the club.
   He's small, he's plain;
   And I love his name.
It causes no hubbub.

THE MUSCULAR MUSKELLUNGE
He's "Muskie" in the literature,
To fishermen in particular.
  Why is this so?
  I think I know:
He's huge and very muskie-ular.

THE PIRATE PERCH
Its urogenital pore
Moved from the aft to the fore;
  To release gametes
  Into fast-flowing creeks
In a way never seen before.

CLEANER FISH
Each Wrasse behaves like a Trouper,
When tending its favorite Grouper.
  It eats parasites
  From various sites.
For both parties, it works out super!

THE SARGEANT MAJOR FISH
Ensign, Lieutenant, Corporal,
Cadet, Captain, or Colonel?
  Where does this guy rank
  In a Private fish tank?
He's a Sargeant Major in General!

A JIMMY DURANTE NOSE
After years of his wife's urgin'
He asked a plastic surgeon:
  "Though this may sound dumb
  Can you fix my rostrum?".
"No, that's normal for a Sturgeon!"

GOBY TWEEN
These fishers get highly paid
For each coral reef they raid.
  Be it Tang Fish or Goby,
  They're really the go-be-
Tweens in the aquarium trade.

THE FISH THAT ISN'T

The Sheepshead
Is not a Sheep's head.
  Nor does it look like one,
  So, it might be quite fun…
To think of another name instead!

THE GAG GROUPER

An aberrant meal of Grouper
Nearly threw me for a looper.
  I almost had to gag
  In a doggie bag
When my supper wasn't super!

THE JACK

Many Jacks live in the sea,
At least presumably.
  But it's hard to know,
  Because they all go
By the same surname, exactly.

THE COD

An Atlantic Cod,
Hooked on a rod
  Puts up a good fight.
  Yes, there's some might…
Packed into his bod!

A GRUNT GRUNTS

The Grunt has a blunt name,
Here's how that became:
  Is he what he says?
  The answer is "yes".
They both are one-and-the-same!

A CROAKER CROAKS

The Atlantic Croaker
Is indeed a croaker.
  To make her stop,
  You've really got
To cag and nearly choke her!

THE SNOOK

I love how hard a Snook
Fights when on the hook.
    But don't snicker,
    Here's the kicker:
He's also great to cook!

THE SOLE

A fish that's called a Sole
Is flatter than a shoe's sole.
    With his brain likely squished,
    It can only be wished…
That perhaps he has a soul!

THE HAKE FISH

Just for the sake,
Consider the Hake
    Might this little cuss
    Be smarter than us?
He's never made a mistake!

LAYING DOWN THE LORE

White-meat Albacore
Rates a real high score…
    In sashimi and sushi:
    Seemingly not mushi.
At least, that's the lore!

IN A BAD PREDICAMENT

"Let me mull it",
Thought the Mullet.
    He was in a mess,
    In total darkness,
Inside a Heron's gullet!

THE MORAY EEL

An eel-like fish called the Moray
Loves to go out and foray;
    Flashing his teeth
    But deep underneath
He's shy; at least that's his story!

TROUT AND SALMON

Let's hear it for the Trout,
Whose praises we should shout.
And Salmon too,
Because the two...
Are cousins without doubt.

THE SHAD FAMILY

The American Shad
Can be easily had;
Just cast a net,
And you might get
A hundred or so: not bad!

THE GRUNION

Grunions nest high on a beach,
As far up as they can reach.
It's instinctual,
Not instructional.
Nothing that they must teach.

THE GAR FAMILY

How many species of Gar
Are known to science thus far?
"Oh, Good Heaven!;
About seven?"
Correct! You win a cigar!

THE STRIPED MARLIN

The fishing line was knarlin'
From the leaps of this lovely darlin'
Such height in the air;
We just didn't care...
Whether it was Sailfish or Marlin.

SPECIAL TAKE-OUT

The Anchovy, usually,
Lives way out in the sea.
But when we dine,
He tastes just fine...
After pizza delivery!

IS THE MACKEREL CATHOLIC?

Here's a Cardinal confession,
About the Churching profession;
   Pope, Minister, Priest, Bishop,
   Deacon, Pastor, and Archbishop,
Holy Mackerel!, what a procession!

SKATE OR RAY?

Is it a Skate or a Ray?
It's rather hard to say.
   They're both very flat,
   And other than that,
It could go either way!

THE WHALE SHARK

This big fellow is truly a Shark:
Not a Whale; the difference is stark.
   His vertical tail
   Helps tell the tale
As a visible field mark.

ON THE FISH FARM

Old McDonald had a farm, E.I.E.I.O.
And on this farm he had some Fish, E.I.E.I.O.
   What Fish did he raise?
   Ones that rate praise:
Mostly Pompano! Oh!

EYELESS CAVEFISH

They're unpigmented and blind;
Really quite one-of-a-kind!
   But their other senses,
   Make recompenses,
So, mates and food they can find!

PLATYFISH AND SWORDTAILS

A fish known as the Swordtail
Has a sword at the end of his tail.
   Platy girls also like
   This lengthy spike,
Which can lead to more sordid tales!

MEMORIALIZED

The Shark couldn't take any more 'a...
His pesty flock of Remora.
  So, he ate the fish
  To fulfill his wish:
I'll have absolutely no more-a-'ya!

CAMOUFLAGED SARGASSUM FISH

In seaweed they tightly stick,
Often looking deathly sick.
  And if you sass 'em,
  In their Sargassum
They'll disappear mighty quick!

THE WAHOO FISH

When you're a Wahoo,
Everyone loves you.
  On the hook,
  Or to cook;
Sportsmen and diners say "Yahoo!"

ON FISHING IN GENERAL

I lured one little stinker
But he was just a dinker.
  So, I had to wait
  For a lunker to take...
My lure— hook, line, and sinker!

A DEEP THINKER IN THE HOLDING TANK

Nothing to do but wait!
Destiny was his fate.
  Might he be set free?
  To be or not to be?
A Fish, or merely cut bait?

BELLY-UP TO THE SANDBAR

So drunk he could hardly think,
But he still had his sharp instinct:
  "Nitrogen narcosis!"
  Was his diagnosis;
As he fell back into the drink!

LET'S PLAY JEOPARDY
It was a Plymouth car.
It travels wide and far.
    It inhabits
    Sea habitats.
Question: What is a Barracuda?

A FAMOUS FOSSIL
Tiktaalik, a non-missing link,
Was precisely on the brink…
    Between Fish and Tetrapod.
    It's been called a "Fishapod".
Which is pretty cool, don't you think?

The Life of a Fish

INTRODUCTION

If you read the previous chapter,
I hope you got some laughter;
    But now we will switch
    To hear the story of a fish.
So, that's what we'll be after.

FRESHWATER VERSUS SALTWATER

Let's start with a basic distinction,
Regarding their jurisdiction:
    Fish in mostly freshwater,
    Versus those in saltwater…
Is a very key contradistinction!

FISH WHEREABOUTS

This issue of residential,
In fish is very essential.
    Freshwater or marine,
    Or perhaps in-between,
Is far from inconsequential!

DIFFERENT ATTITUDES

For an Eel, the ideal strategy,
Is to grow up in streams, mate at sea.
    But to spawn up a stream,
    Is a Salmon's sweet dream.
Talk about reverse psychology!

J.C. Avise, *From Aardvarks to Zooxanthellae*,
https://doi.org/10.1007/978-3-319-71625-1_6

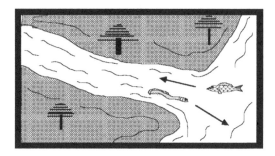

DUAL LIFESTYLES
The Eels are called "catadromous",
While Salmon are "anadromous".
   But either way,
   It's correct to say…
Both species are "diadromous".

EXTERNAL FERTILIZATION
Most fish that live in the sea,
Spawn very promiscuously:
   Eggs from the females
   And sperm from the males
Are released simultaneously!

FISH SCHOOLING
Why fish enter schools is a mystery,
It probably offers some security.
   But the odds can't be beaten…
   That some will still be eaten.
So, it's much like joining a lottery!

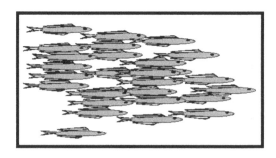

MASS SPAWNING

Sometimes what you see
Is a school acting crazily:
   All the fish ejaculating
   In a climactic mating.
It's a huge spectacular orgy!

FECUNDITY

Fish lay eggs by the million,
Collectively by the billion.
   They must do so to thrive,
   And keep their lineage alive…
'Cause the odds are one in a million (or billion!).

THE PLANKTONIC PHASE

After each fertilization,
Zygotes drift in the plankton;
   Across the sea,
   Very passively;
With little or no sanction.

LARVAL DEVELOPMENT

Each egg then becomes a larva,
Which continues to develop farther;
   But there's little shift
   In the way they drift,
Both growing and going further.

DURATION IN THE PLANKTON

This life-phase in called "pelagic",
And can range from long to quick;
   From just a day or two…
   Up to weeks quite a few.
Let's hope they don't get sea-sick!

LARVAL SETTLEMENT

Then one day, eventually,
They settle down naturally,
   To begin to grow…
   'Mo and 'mo;
Getting ever more Fish-ery!

REEF FISH

Many larvae settle on a reef,
Which must be quite a relief!
  That's why reefs are so rich
  With immense pools of fish;
At least, that's the general belief!

PAIR SPAWNING

Some other fish in the ocean,
Don't like that kind of commotion;
  They spawn separately,
  Two-by-two (or sometimes three!);
In a closer expression of devotion!

SEARCHING FOR PARITY

Fish that lay eggs are oviparous,
Those that give birth are viviparous;
  And some young are born live
  After hatching inside…
A mother who's ovoviviparous.

NEST-TENDING

Many fish tend fry in a nest;
Males typically do this best.
  He may eat one or two,
  (Or perhaps quite a few);
But survival's enhanced for the rest!

FISH NESTS

Most use a depression or scrape
In some body-of-water's substrate,
  Or abandoned shells…
  Of mussels or snails.
Fish employ whatever it takes!

MOUTH-BROODING

The logic is simple to follow:
The mouth is like a big hollow;
  So, some fish brood their young
  Right along-side their tongue;
But they must be sure not to swallow!

INTERNAL FERTILIZATION

Some males can impregnate…
Their mate (or maybe a date!);
How do sperm get in?
Via a modified fin,
And a sort of ejaculate!

FEMALE PREGNANCY

Some fish called "livebearers",
Are superb embryo caretakers.
As in human females,
Pregnancy is entailed…
And there's little room for errors!

MALE PREGNANCY IN SEAHORSES

Whenever his pouch has a vacancy,
A seahorse starts a new pregnancy.
His fatherly role
Is to carry each foal…
Until birthing it out in the open sea!

MALE PREGNANCY IN PIPEFISHES

Pipefish also have male pregnancy;
They too are in the "Syngnathidae".
Each male's portly brood pouch
(And I've got to say "Ouch!")…
Holds young: about twenty or thirty!

FRESHWATER FISHES

Now let's move to a stream or a lake,
Where a different path most fish take.
Most spawn in a shallow nest,
Which a male does his best…
To advertise, tend, and decorate!

NEST-TENDING BY MALES

Each female is a brief guest,
Laying eggs in a male's nest,
After which,
In a sort of "role switch"
The male takes care of the rest.

ATTRACTING FEMALES

One tactic that males employ
Is to use "eggs" as a decoy.
  To attract more dates
  And, thus, more mates…
To his nest-like "incubatory".

EGG MIMICRY IN DARTER FISH

These males have some body part…
That evolved like a fine work of art.
  These egg look-alikes
  Are what a female likes…
She's attracted right from the start!

BOURGEOIS OR "SHOPKEEPER" MALES

For the main nest attendant,
Most of his time is spent…
  Guarding, protecting,
  Patrolling, and aerating,
There's not much to do that he can't!

SNEAKER MALES

During each spawning episode,
As the bourgeois is set to explode,
  A younger fish might sneak
  To get quite a lecherous peek
And likewise ejaculate his load!

SATELLITE MALES

Males of yet another type—
Each called a "satellite"—
  Is a female impersonator,
  Designed by our Creator.
(S)he is really quite a sight!

ILLEGITIMATE FRY

Each brood typically contains
Some fry with other surnames…
  Than the bourgeois male
  So, this had to entail…
"Cuckokldry" as part of the game!

FRESHWATER DEVELOPMENT

There's no true "pelagic phase",
At the baby or juvenile lifestage.
    They just start to grow,
    And before you know,
They're bigger, and greater in age!

FISH FRY

The babies are called "fry",
Although I'm not sure why.
    At this early lifestage,
    They're too young in age…
To bake, or broil, or fry!

INDETERMINATE GROWTH

Most fish just grow forever,
Be it Pike or Perch, or whatever;
    OK, not exactly "forever";
    It depends on whether…
They don't die or get eaten ever!

LIFESPAN

This can vary very tremendously,
From one year, or two, or three…
    To a century or more.
    It's hard to keep score;
Fish always live dangerously!

MARINE FISHERIES

The sea-fishing industry
Often fishes haphazardly.
    Much of their catch
    Is called "by-catch"…
And is tossed back dead in the sea!

FRESHWATER FISHING

Much time and effort is spent…
On every Bass tournament.
    The winners get prizes…
    Of astounding sizes.
But the fish get nary a cent!

FISHING GEAR

Be it a net, seine, or trawl,
All try to catch a big haul;
And, of course, fish must look…
For each camouflaged hook;
They must watch out for them all!

FISH DEATH

When it's time to close the book,
It's often by way of the hook;
But fish also die other ways:
Typically ending their days,
Inside another fish's stomach!

THE END OF THE STORY

That's the story, from start to finish
Of creatures that are so fin-nish;
Husbands, children, and wives,
Yes, they do have full lives…
But it's not quite what I would wish!

Entomology

ON BEEING

Behold the Bumblebee.
Humble as can be.
   She's made to buzz;
    Why? because,
Some things are meant to be.

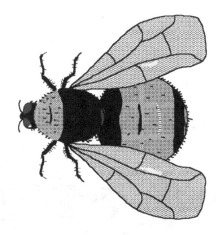

CROOKED BEE-LINE

"Honeybee, won't you be mine?
Fly with me, things will be fine."
   Was the drone being straight
   With his prospective mate?
Or just giving a Honeybee line?

J.C. Avise, *From Aardvarks to Zooxanthellae*,
https://doi.org/10.1007/978-3-319-71625-1_7

THE DRONE DOESN'T GO ON
A Drone leads a simple life
His only goal: find a wife.
   He mates with a Queen,
   Then is ne'er again seen.
He dies happily, free of strife!

THE HONEYBEE DANCE
"Will you come with me to the dance?"
Asked a honeybee, thoughts on romance.
   But the partner said" No,
   With you I won't go,
You're my sister, we're sterile; no chance!"

MINIMUM WAGE JOBS
In the hive of a Honeybee,
Every worker is a she;
   Making honey
   For little money,
As busy as can be.

YOU'VE GOT A POINT
A Hornet is quite an enigma:
We're very afraid of its stigma.
   It's a logo of sorts,
   And a mascot in sports,
But the Insect is what's gonna' sting 'ya!

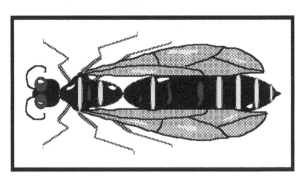

CLEANING UP LEFTOVERS
Dung Beetles, when gathered in herds…
Of droppings, don't mince their words.
   From their dinner stools,
   We hear thru their drools:
"Please pass the seconds and turds."

TRILLING EVENT

Quiet grubs of Cicada persuasion,
Periodically stage an occasion.
Every seventeen years,
They make up for arrears,
And emerge in a raucous invasion.

SOUNDS SIMILAR

Both Cicadas' and Katydids' cadent
Is a trill that neither can patent.
Can't tell them apart?
Well, here's a start:
What Cicada did, Katydid didn't!

WOOD-BORING TERMITES

Their soldiers do the warring,
Their alars do the soaring.
And workers do fine
Keeping kids in line.
Still, their lives are primarily boring!

BUGS OR BEETLES?

Is it a Beetle or Bug?
Through literature I've dug.
But it's never clear,
(At least to my ear);
So, all I can do is shrug!

MUTE POINT

Silent is the Butterfly.
Without a sound it flutters by.
It's very cute,
But stone cold mute--
No vocal cords to utter by!

LICE

Up close, their mouthparts actually,
Are an impressive set of toolery,
To pierce, suck, and crunch,
Chomp, nibble, and munch;
In short, to help them act chewily.

MOSQUITOES

Swatting mosquitoes;
On and on it goes.
   These pesty pests
   Give us little rest.
Not even to doze!

THE GNAT

The gnat
Knows where it's at.
   She bites a vein
   Causing pain.
And that's that!

FLEAS

Perhaps we can all agree,
That we abhor the Flea.
   To keep them at bay,
   There's more than one way:
Either fight or flee!

IS THAT A WALKINGSTICK?

Let's cut to the quick.
Is it Insect or stick?
   Hard to say
   Either way,
Take your pick!

INSECT ONTOGENY

Egg, larva, pupa, adult,
Is one development route
   But another pathway
   Puts nymphs into play.
That's what Insect life is about!

KUNG PHOOEY

"Listen carefully, Grasshopper,
For I have wisdom to proffer."
   Said the old Kung Fu sage
   To the green nymphal stage,
Who was far too young for the offer!

LET'S PLAY JEOPARDY

It chirps in a thicket,
But a sport is its ticket...
  To great acclaim,
  And lasting fame.
Question: What is Cricket?

ANTS

Ants are extremely fleet,
It's quite an amazing feat,
  Never tripping or stumbling
  It leaves me wondering
How they synchronize all six feet!

ON THE FARM

"I guess it won't do no harm",
Said his Mom without alarm.
  She had just told
  Her six-year-old,
"Yes, you can have an Ant-farm!"

A FLY IS WHAT HE DOES

The name for this Dipteran guy
Serves double-duty, here's why:
  There's a perfect conjunction
  'Twixt form and function...
Conveyed in a single word: "Fly".

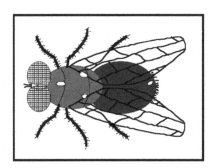

LET'S PLAY JEOPARDY

This cute car merits a hug;
As an Insect, merely a shrug;
  It's made by Volkswagon,
  But is not its station wagon.
Question: What is a Beetle or Bug?

ON THE FARM
   After being repeatedly stung,
   He yelled at the top of his lung:
      "Oh, behave,
      You behive!"
   As his smoker he faster swung!

Invertebrate Zoology (other than insects)

CALAMITY

Other Clams in her bed tend to crowd her.
She can't even go out for a powder.
    Her best chance for escape?
    A fisherman's rake.
But then she would end up in chowder!

PEARLS OF WISDOM

A pearl is a jewel resplendent,
Prized by royalty, bourgeois, and peasant.
    From an Oyster's perspective,
    Which is far more objective,
It's an adenoma nacrescent!

AN OYSTER BED

In bed, they constantly lay;
All night but also all day.
    Do they need this much rest?
    Or are they depressed?
It's rather hard to say.

INTERTIDAL MUSSELS

How Mussels hold firm to a rock,
Might come as a bit of a shock.
    With waves or other tustles,
    They don't flex their muscles;
But instead use glue for a lock.

© Springer International Publishing AG 2017
J.C. Avise, *From Aardvarks to Zooxanthellae*,
https://doi.org/10.1007/978-3-319-71625-1_8

ANOTHER BIVALVE

A shell simply known as the Ark,
Lies buried in mud in the dark.
    Fossils show
    That it's always been so:
Ever since Noah's ark!

YET ANOTHER BIVALVE

This cutie is the Scallop,
Whose muscle packs a wallop.
    When it snaps its shell,
    It "swims" quite well.
Though hardly at a gallop!

ABALONE ALIBI

The tasty sea abalone…
Just wants to be left alone. He…
    Replied with a glimmer
    When I asked him to dinner:
"No thanks, I've a date"— Ah, baloney!

LITTORAL ZONE

Marine beaches are littered fully.
Tons of critters, literally.
    Starfish and snails,
    And clams by the pails…
All doing their things littorally!

WISHING UPON THE HEAVENS

Let me quickly explain
Why a Starfish is poorly named.
    Not a star nor a fish;
    But if you really wish:
"Asteroidea" can be retained.

HIKING

A Centipede spent a whole week,
To hike to the top of Pike's Peak.
   At the end of the trip
   He invented the quip:
Thrill of vict'ry, agony of da feet!

STRANDED

The Man-O-War mumbled, "Oh damn!"
Too close to the beach he had swam.
   With a Woman-O-War,
   They'd been washed high ashore.
Now they're both in a Jellyfish jam.

CTENOPHORES

The Ctenophore or Comb Jelly
Often ends in a Sea Turtle's belly.
   But wouldn't it be neat
   If the Turtle could eat
Peanut butter and bread with the Jelly?

ANEMONIES MANY ENEMIES

Whenever a Sea Anemone
Happens to see an enemy,
   By waving her arms
   She issues alarms:
"You better not mess with any o' me!"

AN ANEMONE STING OPERATION

She said, "Come here, I insist."
Her invite was hard to resist;
   The reef fish soon found
   He was tentacle bound
By the sting of her nematocyst.

GOING ALL OUT ON LIMBS

Daddy Longlegs are somewhat slimmer,
Fitter, and physically trimmer...
  Than any insects;
  And with eight legs, not six,
They are also somewhat more limber!

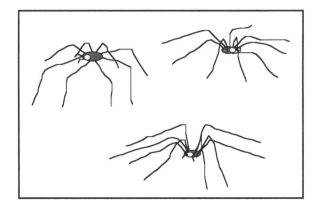

WEB OF LIES

Spider females-- how they deceive!
They eat their mates, then "bereave."
  When police nose around,
  She claims: "He's out of town."
What tangled webs they can weave!

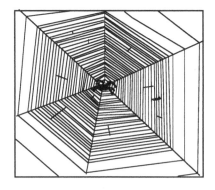

SEA SPIDERS

These creatures live in the sea,
Where the males act fatherly...
  By carrying eggs
  On specialized legs.
It's called "external male pregnancy".

HOMELY AND BEAUTIFUL

The Snail became very tired,
Making his shell so spired.
  Building his tower
  Took many an hour;
But he finally got in-spired!

THE BAHAMIAN CONCH

A huge snail called the Conch,
(Which happens to rhyme with "bonk"),
  Lives all alone
  In his beautiful home.
Cloistered like a monk.

GIANT SQUID

Amazing creatures they are:
Truly spectacular!
  But they're scary too,
  To me and to you;
They're so darn tentacular!

OTHER SQUID

Some Squid
Keep well hid
  By squirting ink,
  And in a wink;
Predators they rid!

OCTOPUSES

How many tentacles occupy
A squad of nine total Octopi?
  Seventy-two;
  Here's how I knew:
Nine times eight I multiply.

ANOTHER CEPHALOPOD
It fills an ecological void
That other mollusks avoid.
    Living deep in the sea,
    It swims backwardly;
It's the nautical Nautiloid.

HE DESTROYED MY LETTUCE PATCH!
A Slug is a shell-less Snail
Who really belongs in jail.
    The scene of his crime
    Was covered in slime
As was his shiny trail!

THE BEAUTIFUL NUDIBRANCH
The end of her name rhymes with "bank",
And she's gorgeous, to be frank:
    Though she crawls around nude,
    It's never rude.
To stare at a Nudibranch.

LAZY NO-GOOD SPONGES
They're lethargic like no other;
To move they don't even bother;
    And you can bet…
    They're always in debt,
From sponging off one another!

A MACHO LOBSTER
I'm fit, I'm hardy, I'm hail,
Almost as tough as a nail.
    I'm a sexy beast,
    To say the least;
Yes, everyone wants my tail!

ON A SEAFOOD FARM
A pampered hand-raised Shrimp,
Is really quite a wimp.
    When trapped in a net,
    And still soaking wet,
He quickly tends to go limp!

GOING IN FOR THE KRILL

When feeding on Krill,
Each Whale may kill
   'Bout a million or more.
   But whatever the score,
It's enough to get his fill.

THE SEA CUCUMBER

This Echinoderm is a wonder
Who lives about 30 feet under…
   The surface of the sea,
   Where you can plainly see…
It looks like a normal cucumber!

ANOTHER ECHINODERM

The beachcomber let out a holler,
When he spied a silver dollar.
   But his buzz didn't last,
   So, he just walked on past;
What turned out to be a Sand Dollar.

A SEA URCHIN EMERGENCY

The surf was really a'surgin,
Pounding the helpless Urchin.
   Though it broke a few spines,
   He didn't much mind.
There was clearly no need for a surgeon!

LOWLY SEA SQUIRTS

Regarding Sea Squirts or Tunicates,
Their dorsal notochord indicates…
   Quite bluntly stated
   They must be related
To all of us higher Vertebrates!

TAPEWORMS, EARTHWORMS, ETC.

There are many kinds of Worm:
From flat to round, squishy to firm;
   Parasitic or free-living,
   They all keep giving…
Me shivers and making me squirm!

POLYCHAETE WORMS
Some polychaetes have many "feet",
And can be very fleet.
　Others are sedentary,
　And must be more wary…
Of all predators they meet!

BLOOD-SUCKING LEACHES
From streams to sandy beaches,
You sometimes encounter Leeches.
　It's not your fault,
　Just rub on some salt.
And pretty soon they releases.

A TICKLESS SITUATION
The bite of a Tick
Can make you sick
　With Lyme disease.
　So, you must seize…
Any chance to kill him quick.

THE MITE
The tiny Mite
Loves to bite;
　Day or night
　At any height.
That's not right!

GETTING OLDER AT A SNAIL'S PACE
Can we age a Periwinkle…
In an eye's twinkle?
　Approximately, "Yes".
　Here's the best guess:
About one year per shell wrinkle.

DON'T GET PINCHED
Whenever you try to grab,
Almost any species of Crab;
　It lays down the law
　With a nip from its claw;
You get pinched before the nab.

THE HERMIT CRAB

Along the shore he can roam,
In a Snail's abandoned dome.
  It's owned free and clean,
  'Cause the bank holds no lien,
On his mobile adopted home.

FIDDLER CRAB

This poem is just a spoof,
For which there is little proof.
  The Crab caught a ride
  On a really high tide;
And became "fiddler on the roof."

HORSESHOE CRABS: "LIVING FOSSILS"

Think of all their trials and tears,
All their hopes, their dreams, their fears;
  Though they didn't plan it,
  They've lived on the planet
For 200 million years!

BARNACLE BILLS

You can search as hard as you will
To find even one Barnacle...
  Named Tom, Dick, or Harry,
  Moe, Curly, or Larry.
Each one has got the name Bill!

PODCAST

Isopod and Amphipod;
Arthropod and Gastropod;
  Decapod, Copepod,
  Isn't it really odd...
That "pods" were so favored by God!

WHOOPS

He gave his mother a tug:
"I think I need a hug."
  Something odd slipped in...
  My medicine.
I swallowed a round Pill Bug!

TRUE MYTHS

Here's some oxymorons I know:
Honest Cheetah and unhip Hippo.
    But even more impish
    Is one more shrimpish:
"Jumbo Shrimp" wins Best of Show!

A SURE BET AT THE RACETRACK

In a race of a Centipede
Against a Millipede,
    I'd be all set
    To place my bet
On the one with greater speed!

THE MILLEPEDE

Coordinating with each other
Must generate quite a bother.
    How can so many feet
    Not skip a single beat?
One step just leads to another!

THE EARTHWORM

He slept late, in his earthen berm,
Moist and cozy and warm…
    Until afternoon most days,
    'Cause he knew the phrase:
The early bird gets the worm!

ON THE FARM

Oysters are also farmed
But don't be too alarmed;
    Such aquaculture
    Is like agriculture:
A crop's life is charmed.

AQUACULTURE

This quite sea-sonal activity
Raises creatures in captivity.
    From fresh Oysters to Fishes,
    They end up on your dishes,
Because eating is your proclivity!

Microbiology

CALCULATING MICROBES

Bacterial numbers are high.
On statistics they rely.
  Mathematically,
  With bountiful glee,
They divide and multiply.

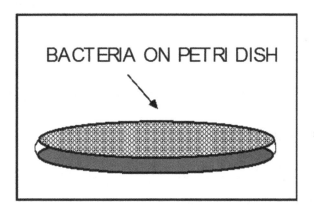

MOULDED

Some very strange creatures among us,
Crave bread, causing problems humongous.
  God broke the mold,
  So I've been told,
When he invented the fungus.

© Springer International Publishing AG 2017
J.C. Avise, *From Aardvarks to Zooxanthellae*,
https://doi.org/10.1007/978-3-319-71625-1_9

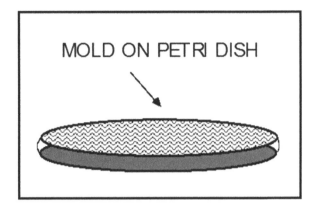

GESUNDHEIT

Do you know why people sneeze?
So that germs disperse with ease.
  It evolved for the virus,
  Rather than us;
As their way of spreading disease.

YEAST

The Yeast
Is a teeny beast;
  Very small
  That's all.
To say the least!

THEY'RE EVERYWHERE

Microbes litter the sea,
That's virtual reality.
  Even pristine waters
  Hold hordes of these squatters.
But they're mostly too small to see!

PLANKTON

How can teeny-tiny Plankton
Weigh vastly more than a ton?
  It's cumulative,
  And the sea is alive,
With billions and billions of 'em.

VERTICAL MIGRATION

Most Plankton in the sea
Can't swim very actively.
  So, rather than roam,
  They float up and down,
By adjusting their buoyancy!

MICROALGAE

Golly gee!
Micro-algae
  Are part of the plankton too
  Who knew?
But now I see.

ZOOXANTHELLAE (SYMBIOTIC "ALGAE")

They live within coral tissues,
Which brings us straight to the issues:
  They both need and feed their hosts;
  Then depart leaving skeletal ghosts.
Coral bleaching and death next ensues!

DYING REEFS

What causes these sad departures?
Mostly rising water temperatures...
  From climatic change,
  And in the exchange...
We're left with poor reef caricatures!

PROTOZOANS

Who could ever be a fan
Of a one-celled Protozoan?
  But a protozoologist
  Is no apologist
For praising them all that she can.

UNICELLULAR PARAMECIA

They immodestly sighed:
"We have skills bona-fide."
  In verbage and math,
  We've evolved on a path...
To conjugate and divide!

OUR MICROBIOME
   I hesitate to discuss
   A topic that raises a fuss.
      That microbes outnumber…
      Our cells, makes me ponder:
   There are more of them than of us!

MORE ABOUT OUR MICROBIOME
   They're everywhere, inside and out;
   Generally, all about.
      Yes, our bodies are rife
      With microbial life…
   With untold medical clout.

SAGE WORD USAGE
   "Bacterium" or "Bacteria";
   What is (or are) the criteria?
      Plural or singular?
      A problem that's similar:
   Cafeterium or cafeteria?

MORE SEMANTIC ANTICS
   Protozoa or Protozoans?
   Paramecia or Parameciums?
      Funguses or Fungi?
      If it was I (or we),
   I'd (or we'd) ask microbiologians!

DEADLY GERMS
   When docs and nurses gather,
   They agree that they would rather
      Use an antibiotic
      For anyone sick
   Than end up with a cadaver!

Botany

LEAVING OUT BOTANY

I've tried to branch out, but can't do it.
Things herbaceous I just can't intuit.
  If I tried to write poems
  Of xylems and phloems,
My style would get stigma stuck to it.

J.C. Avise, *From Aardvarks to Zooxanthellae*,
https://doi.org/10.1007/978-3-319-71625-1_10

Anatomy, Physiology, and Medicine

WE'RE ALL WET (75% WATER)
From a business vantage it's tacit:
Our bodies have many a facet.
  They're our piece of the rock,
  Interest, capital, stock,
And our primary liquid asset!

HONESTY IS THE BEST POLICY
I admit, I can't tell a tibia
From clavicles or a mandibula.
  So, to this very day,
  I can honestly say
That never have I told a fibula!

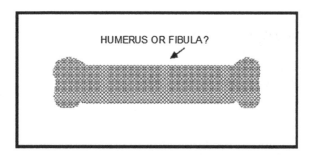

HUMERUS OR FIBULA?

NOT FUNNY

Although jokes about it are numerous,
It wasn't put there to humor us.
  It's no laughing matter
  Whenever you batter
The funny-bone end of your humerus!

GLUTEUS MAXIMI

These portable pillows poke
Far aft, on many folk.
  Standing or sitting,
  They're always fitting
As butts for many a joke.

NO BED OF ROSES FOR ME

A rose is a rose is a rose.
That's obvious, I suppose.
  But a thorn in my derriere,
  Has made me much warier…
Of where I sit down for repose.

OUT OF CIRCULATION

A tightness in the chest
Feels constrictive at best.
  That's merely the start,
  When clots place the heart…
Under cardiac arrest.

THE HEART

Whether beating slowly or fast,
This organ was built to last…
  To the end of our days
  When the Coroner says,
"By definition, you've passed!"

BLOOD

Whatever your religion, blood is red.
Whatever your race, blood is red
  Whatever your skin, blood is red.
  Whatever your sex, blood is red.
Whenever its gone, you are dead!

PREGNANCY'S ENDOCRINO-LOGIC

A mother and fetus at war
Defies conventional lore,
  But nature's logical…
  Immunological…
Battles are hard to ignore!

VAS DAT?

The convoluted vas deferens,
Is a seminal source of reference…
  For a sperm headed out;
  It's the thoroughfare route,
Whose evolution made a vast difference!

ODE TO AN OVARY

Oh, ovary,
How very…
  Kind of you
  To spew…
An egg per month (almost every!)

THE PANCREAS

A strange little gland, the pancreas;
It sometimes gets cantankerous.
  It secretes juices
  For multiple uses,
And it's terrible when it gets cancerous!

THE LIVER

This organ has more than one way,
To assist a human gourmet.
  Her glucose it releases,
  And another good use is…
As a tasty Goose-liver paté.

IT ONLY STORES BILE

The gall bladder
Doesn't much matter.
  It often gets approval
  For surgical removal,
With hardly an eye's batter!

THE LARYNX

Our "voice box" is the larynx,
Whereas birds have a syrinx.
    They are fore versus aft
    Of the tracheal shaft.
Which thus lies right in-be-twinxt!

A FULL COURSE OF A MEAL

Its path thru the alimentary
Is truly quite elementary.
    Food enters the mouth,
    Meanders down south,
And exits by the rear-entery.

NATURAL RELAXATIVE

When omnivores get in a bind,
They need all the help they can find.
    They change what they eat...
    To fibre from meat,
And soon leave their problems behind!

THE LONG AND THE SHORT OF IT

The human intestine,
Is very interestin'.
    It has small and large parts,
    Releases farts...
And other small indiscretions!

THE RESPIRATORY SYSTEM

Lucifer's doorbell gets rung...
Whenever a convict is hung.
    He dies lacking air,
    Which had come from where?
Obviously, a lung!

THE SKELETAL SYSTEM

Bones and our skeletal structure,
Are our calcified infrastructure.
    When they're not broken,
    We're extremely betoken
To this inner superstructure!

ADRENALS TO THE RESCUE

Whenever we're on the verge…
Of a fighting or fleeing urge,
   An organ obscure
   Kicks into high gear,
And provides an adrenalin surge!

THE SKIN

It's likely our largest organ,
So, it comes at quite a bargain.
   Except when we need
   To profusely bleed,
And it must heal all over again!

99% PERSPIRATION, 1% INSPIRATION

Perspiration, though sometimes a mess,
Is really a key to success.
   So, the birds of our lands
   (Who all lack sweat glands),
Must be doomed to failure, I guess!

EXCRETION

Oust nitrogen: that's the name
Of the excretory game.
   Since urine is watery,
   Kidney stones actually
Cause the excretiating pain!

DEALING WITH THE CARDS YOU'RE DEALT

As the organs above grow older,
Any can go "OUT OF ORDER".
   Depending which one,
   This isn't much fun,
And you might even have to fold'er!

THE BRAIN

The human brain
Can be quite a pain.
   Getting a stroke,
   Is truly no joke;
May it never happen again!

POSITIVELY NEGATIVE THOUGHTS
And then there is depression,
That can become an obsession...
    With negativity
    Which positively...
Recycles to further depression!

PATENTLY DUMB
Did you hear of the silly invention?:
A helmet for headache protection.
    An aspirin, for sure,
    Is an ounce of cure
Well worth this pound of prevention!

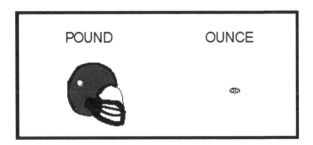

MORE ON THE BRAIN
But there's far more than that,
Going on under your hat.
    Any brain in absentia
    Can become dementia.
After which, that's about that!

THEORY OF THE MIND
Are we conscious of our conscience?
That's mostly a question for science.
    But it's easy to say,
    That neither's in play...
When we become unconscious!

THE STOMACH
All this talk of organs and such,
Is simply way too much.
    Please don't say any more
    About organs galore:
I've had all I can stomach!

IMPOTENCY

The mission of sperm emission
Sometimes goes out of commission.
    The reasons can vary;
    When temporary,
Intromission is on intermission.

VARIOUS DISSECTOMIES

I first had a tonsillectomy,
And later an appendectomy.
    These worked out fine,
    But I drew the line
When my doctor proposed a vasectomy!

IT ISN'T FUNNY, DOC!

During operational glitches
Some surgeons exhibit mood switches
    To a more joking air.
    What do they care?
They chuckle, we end up in stitches!

MAKING A SPECTACLE FOR HIMSELF

At first, he was rather skeptical,
But soon he saw need for a spectacle.
    The optometrist
    Was an optimist:
"Perhaps you'll look more respectable."

VICE ADVICE

Some people crave smoking sensations,
While others need excess libations.
    No-one seems too alarmed
    That his health might get harmed.
It's enough to make doctors lose patients!

A FLABBERGASTING PROGNOSIS

His gut was like jello, not firmy;
So, the doctor lectured him sternly:
    "You'd better get wise
    And exercise
Or you'll end up in an in-firm-ary!"

HOSPITALS

Whenever you have a calamity,
Docs and nurses are nice, in reality.
   Still I'd greatly prefer…
   Not to have to endure…
Any of their hospitality!

THE V.A. HOSPITAL

The student made a big mistake
In the career he decided to take.
   A hospital for Veterans?
   I'll become a veterinarian!
I've seen errors, but that takes the cake!

ON THE FARM

Now enter the veterinarian,
Whose clinic is more agrarian;
   Unlike M.D.'s,
   He treats many species.
He's a medical egalitarian!

A COMPLICATED PRACTICE

The typical veterinarian
Treats as many lives as he can.
   From Birds to Cattle,
   It's always a battle:
From a Horseman to a Canary-an!

HIPPOCRATES' LEGACIES

I hate to be hypercritical
About the oath hippocratical.
   But, each outrageous fee
   From the doctor I see…
Makes her pledge just seem hypocritical!

THE END OF THE LINE

Then one day you'll discover,
That you seem to have "crossed over".
   All your pain is gone,
   Because you've "moved on".
(Unless, of course, you recover!)

FOR HEAVEN SAKE, WHAT'S THE SOUL?

Do we really have a soul?
And if so, what is its role?
   Does it carry on
   After we're gone?
And what's its ultimate goal?

Ecology

THE SCOPE OF ECOLOGY
Defined most broadly,
The field of ecology…
 Deals with every environment,
 And species that therein rent…
Every home to raise every family.

LIFE IS A CHESS GAME
Bishop, Knight, King,
Castle, Rook, Queen.
 But in life's game,
 We're all the same:
Mere Pawns in some broader scheme!

NICHE
Some pronounce nitch like "neesch";
Others rhyme the word with "hitch".
 By whatever name,
 It remains the same:
An important concept to teach (or tich)!

MORE ON NICHE
A niche is an ecological role:
Each species "heart and soul".
 Where it lives, what it does,
 All in the cause…
Of survival, its ultimate goal.

© Springer International Publishing AG 2017
J.C. Avise, *From Aardvarks to Zooxanthellae*,
https://doi.org/10.1007/978-3-319-71625-1_12

EMPTY NICHE

Can a niche ever be empty,
At least predominantly?
    Some say "yes", some say "no".
    Even though…
Some niches hold species a'plenty.

NULL AND AVOID

Conversely, can a niche be full?
This query is likewise dull.
    How many sinful sins…
    Dance on heads of pins?
Such questions should be voided and null!

AN ECOLOGICAL VACUUM

Are we then to presume
That an ecological vacuum
    Can exist or cannot?
    I don't care a lot!
In either case, let's resume.

HABITAT

Each class of habitat
Is usually where it's at.
    The type of place
    Called home base:
Where a species hangs its hat.

SYMBIOSIS

We've searched for a diagnosis
On the meaning of symbiosis.
    Commensalism plus mutualism?
    And even parasitism?
No answer is under our noses!

FOOD WEB

The phrase "food web"
Is now used instead
    Of "food chain"
    Which became
Essentially dead.

FOOD PYRAMID
"Food pyramid"
Is likewise amid…
  Other such words
  Now seldom heard;
Of which we are rid!

CHARASMATIC MEGAFAUNA
Panda, Shark, Zebra, Rhino,
Are spectacular, we all know;
  But then so too…
  Is an Elephant Shrew,
Who likewise merits a "Wow!"

ECOLOGICAL COMMUNITY
How much functional unity
Lies within a biotic community?
  It depends quite a lot
  On to whom you talk.
Take a middle stance for immunity.

ECOSYSTEM
Larger than a community,
It's hard to define definitively:
  All lifeforms, plus other stuff,
  And if that's not enough,
They interact with just partial unity!

KEYSTONE SPECIES
Bald Eagle, Grizzly Bear, Sea Otter,
Grey Wolf, Caribou, and a few other;
  Such species hold the key…
  To ecosystem integrity:
They hold everything together!

GAIA HYPOTHESIS
This theory is quite teleological,
And seems to me highly illogical:
  Communities have a capability
  To regulate Earth's habitability.
Is this science or is it just mystical?

CHANGING ATTITUDES

In this age of climatic change,
We quickly must rearrange...
   Our thoughts on the topic:
   We can't be myopic,
Nor act like we're all deranged!

THE OZONE HOLE

What caused the ozone hole?
Was it burning too much coal?
   No! Chlorofluorocarbons
   When released by the tons.
Were responsible, on the whole.

GREENHOUSE EFFECT

Science has a warning
That truly is alarming:
   Heat in our skies,
   And sea level rise.
Are due to global warming!

GREENHOUSE GASES

It's clearly time to decide
How much carbon dioxide...
   Our atmosphere can take,
   Before we all bake
In self-induced mass suicide!

ADMONITIONS

Let's get off our collective asses,
And emit far fewer such gases...
   Into the atmosphere.
   If we don't, I fear...
Our time on Earth too soon passes!

IN SELF-DENIAL

Despite all these admonitions
About atmospheric emissions,
   The Head of our land
   Has his head in the sand,
Focused on other missions!

CRYSTAL BALL

When into the future we peer,
Perhaps the broadest frontier,
   For life's evolution
   Is to find a solution
To the ecological crises we fear.

THE POPULATION BOMB

Human overpopulation
Is a horrible situation:
   The most basal cause
   Of Earth's many flaws.
It's quite an abomination!

EXTINCTION IS FOREVER

Despite all the jubilation
Regarding God's Creation,
   Species' extinction
   Has one distinction:
It's life's final destination!

Ethology

THE COMMENSAL OF MUTUALISM
One common extended prognosis,
For a parasite and its hostess,
   Is violent hate,
   Then open debate,
Then maybe, friendly symbiosis!

GOOD COURTMANSHIP
Females are frequently chased
By ardent males, in great haste.
   But women can win
   By simply giving in:
They really don't want to stay chaste!

FEMALE CHOICE
Animals choosing a mate,
Prefer more than one date.
   Because females require
   That their childrens' sire,
Be no less than the ulti-mate!

AVIAN LEKKING BEHAVIOR
For species that have a "lek",
The males display like heck…
   In a gathering place,
   Vying face-to-face,
All the better for females to check!

© Springer International Publishing AG 2017
J.C. Avise, *From Aardvarks to Zooxanthellae*,
https://doi.org/10.1007/978-3-319-71625-1_13

MORE ON LEKS
These communal male displays…
Can occur in several ways:
    Grouse dance to seek romance;
    Grackles pose in an upright stance;
While all the Hens closely appraise!

MATING RITUALS
Hummingbirds zoom real high,
Straight up into the sky;
    Then dive back down
    To near the ground;
All under a mate's watchful eye!

BOWERBIRDS
In Australia, these males build a bower…
In order to really wow her.
    If she likes what he's done,
    They're in for more fun…
When he really gets to "know her!"

SPERM COMPETITION
After all the matings are done,
Then comes fertilization;
    All sperm cells compete
    For each chance to meet…
Any not-yet fertilized ovum!

WOMEN'S LIBERATION VIA PARTHENOGENESIS
Some lizards view males as nemeses;
Bums to be kicked off the premises.
    Such females on Earth
    Can give virgin birth
By engaging in parthenogeneses.

HUMAN PARTHENOGENESIS
Humans also had parthenogenesis,
In a Biblical version of genesis:
    The Virgin Mary gave birth…
    To Jesus, right here on Earth,
Through a Heavenly stroke of genius!

ASEXUALITY

This truth is actually factual:
Some species are strictly asexual;
   With no sex altogether,
   It makes us ask whether...
They really enjoy their life at all!

HERMAPHRODITISM

Behold the hermaphrodite:
A part male, part female sight.
   Who are we to say
   If (s)he's straight or gay?
We might be half wrong and half right!

CUCKOLDRY

A wife acts illicitly bold;
Very calculating and cold.
   Her vows on the altar,
   She chooses to alter:
"I promise to have and (cuck)hold.

TO THINE OWN SELF BE TRUE

Snails may be able to theorize...
A great deal more than we realize.
   Their mating perspective,
   Is quite introspective:
To "know thyself" means self-fertilize!

PRUDENT PREDATION

A Morel applies to this story
Of gastronomical glory.
   The wisest of eaters
   Avoid Amanitas
When munching fungal quarry.

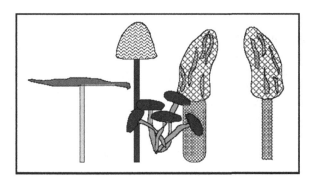

MAMMALIAN DIETS

His food is delicious,
Healthy and nutritious;
   A Humpback Whale
   Feels hearty and hail,
After swallowing schools of Fishes!

MORE ON MAMMALIAN DIETS

Some Orcas or Killer Whales…
Also eat fish by the pails;
   But here's the deal
   Others also eat Seal;
This is what science unveils!

MORE ON MAMMALIAN DIETS

The dietary habit
Of every Rabbit,
   Is vegetarian,
   Yet egalitarian,
In its garden habitat!

MORE ON MAMMALIAN DIETS

Other mammals are omnivorous:
Both herbivorous and carnivorous.
   From Bugs, Worms, and Beetles,
   To leaves, grass, and needles;
They can eat far more than any of us!

MORE ON MAMMALIAN DIETS

Depending on what they eat,
Ranging from lettuce to meat,
   Most carnivores
   Eat omnivores…
Whenever the twain shall meet!

AVIAN DIETS

Bird diets are quite varied too,
Eating nearly all that they view;
   But key limits arise,
   Based on food size,
'Cause birds aren't able to chew!

PISCINE DIETS

From insects to other Fish,
Anything is on their dish;
    Even their poop,
    Can make a nice soup;
Fish eat whatever they wish!

MAMMALIAN SONG

A Humpback Whale's song
Goes on for very long;
    To communicate
    And attract a mate;
(Unless he gets it all wrong!).

MORE ON MAMMALIAN SONG

Coyotes mostly yip at the moon;
But in truth they'd rather croon.
    Do you know why...
    They seldom try?
They simply can't carry a tune!

AVIAN SONG

Birds sing for two reasons only:
To defend a territory,
    Or to get a mate.
    (Or pontificate...
Telling everyone their story!).

PISCINE SONG

Fish seldom sing,
But here's the thing:
    Some grunt, croak, or hum,
    That's about the total sum;
So, at least it's sumthing!

FREEWAY TOLLS

To this impact on viability,
Creatures lack adaptability.
    It's squish, th-thump, splat
    For Toad, Turtle, and Cat;
They're no match for auto-mobility!

CAUTION: ANIMAL CROSSING

To help the faunal community…
Cross certain roads with impunity,
These signs of our times
Have varied designs
As aids to auto-immunity.

EMISSIONARIES

High-roading, it seems, is a mission,
For clergy with righteous conviction.
Through many a speech,
They endlessly preach
Their missionary position.

ANIMAL MIGRATION

Long-distance migration
Takes a lot of navigation.
Plus, endurance and fortitude,
So, don't take the attitude…
That they're going on a vacation!

DIAPAUSE

Many species evolved diapause:
An ontogenetic pause…
   In real early development.
   Based on the sentiment…
That it helps the broader cause!

HIBERNATION AND AESTIVATION

To distinguish hibernation
From elongated aestivation,
   Is a challenging chore
   That we mostly ignore.
It causes too much aggravation!

FILIAL CANNIBALISM

Do fish ever eat their own young?
Yes, occasionally this is done;
   Such "filial cannibalism"
   Boosts adult metabolism,
But juvenile profits are none!

MATE-EATING (OR PACKING LIFE IN)

After Spiders and Octopi mate,
She eats her mate! Such is fate.
   It's completely unilateral,
   So, honeymoon and funeral…
Both occur on their first dinner date!

BIOLUMINENCENSE

A thousand leagues under the sea,
It's so dark you simply can't see.
   So, it made perfect sense
   For luminescence
To evolve repeatedly.

PITCH BLACK

For example, Lanternfish
Fluoresce whenever they wish.
   To attract a mate
   Or alleviate…
Blackness that's otherwise pitch!

LESSONS IN MORES

To act with humble humility,
Is an art of social civility.
    So, we try to be modest,
    Those of us blest…
With stupendous ability!

INSTINCTUAL BEHAVIOR

What is an instinct? In fact,
It's far from a science exact.
    How do proteins and genes
    Yield ethereal things…
Like behaviors, emotions, and acts?

CAUSAL CONNECTIONS

The concept of an instinct,
Is more nuanced than you think.
    How material things…
    Drive such deep feelings,
Is a glaringly still-missing link!

CULTURAL TRANSMISSION

When a Primate behaves somehow new,
Others learn and can master it too.
    "Wow, that's cultural gain!"
    The ethologists claim.
I just say, "Monkey see, monkey do!"

WISE BEHAVIOR

In case you hadn't heard,
I guess I'll pass the word:
    Whatever you do,
    Be one of the few…
To drink upstream of the herd!

MORE ON ANIMAL BEHAVIOR

There's much more about behavior,
Little tidbits for you to savor,
    Throughout this book,
    So remain on the look!
Ethologically you'll be the savvier!

Evolution

FATHER KNOWS BEST
Life arose from primordial slime,
Evolution's ladder to climb.
As fossils well show,
The process was slow.
Mother Nature can't fool Father Time!

PALEONTOLOGISTS
These scientists so professorial,
Root hard for each fossil memorial.
So much digging and burrowing,
Might they start worrying:
"Are we becoming fossorial?"

FOSSILS
Each well-preserved fossil
Is a long gone testimonial
To some creature past
Who just didn't pass
Its evolutionary trial.

THE LEGAL PROFESSION
Nature red in tooth and claw,
Is what Darwin often saw;
His clear detection
Of natural selection
Introduced us to natural law.

© Springer International Publishing AG 2017
J.C. Avise, *From Aardvarks to Zooxanthellae*,
https://doi.org/10.1007/978-3-319-71625-1_14

LAWFUL PUNISHMENT

Let me be very succinct:
Man's and nature's rules are distinct.
Break a human decree,
You may not stay free.
Break a natural law, you're extinct!

MICRO- AND MACRO- EVOLUTION

No matter how small the solution,
Any genetic change is evolution.
Except that
We expect that...
Sometimes it's a big revolution!

MUTATION

This is the ultimate source,
Of course, of course,
Of all genetic variation:
The foundation of creation,
By any other evolutionary force.

GENETIC DRIFT

Genetic drift
Is any shift
In gene frequency
Randomly
And that's it!

GENE FLOW

It's the movement of genes
Within any single species...
Via individual dispersal,
With or without rehearsal,
By any traveling means.

PHYLOGENY

To every arborists' glee,
We'll soon have an evolutionary tree...
Showing life's full story,
In all its great glory:
Branches, roots, and canopy!

PHYLOGEOGRAPHY
Biogeography
Teamed with phylogeny,
  Tells how life is dispersed
  Across planet Earth,,,
Now and throughout history!

CAPTIVE BREEDING
"Artificial selection",
Requires the detection
  Of the best phenotypes,
  And, thus, genotypes,
For each new generation.

SPECIATION
When a species splits in two,
What arises is something new,
  Never present before,
  Whereby and therefore,
Life diversifies anew!

ADAPTIVE RADIATION
For species' proliferation,
An adaptive radiation...
  Is an evolutionary feat
  That can hardly be beat:
It's a cause for jubilation!

HYBRIDIZATION
Some hybrids seem rather quirky,
Non-functional and non-worky.
  And their progeny
  Make the family tree
Less branchy and more networky!

INTROGRESSION
This is the movement of genes
Between or among species,
  Via hybridization...
  Without any realization
Of consequences unforeseen.

REPRODUCTIVE BARRIERS

These retard introgression
Via any deep depression
  In a hybrid's fertility
  Or else its viability;
Such animals merit compassion!

AVIAN ENLIGHTENMENT

Birds are smarter than reptiles, much brightened,
Having opted for flight when frightened.
  They evolved feathered zones,
  Air sacs, hollow bones,
And thereby became much enlightened!

MID-LIFE CRISIS

Salamanders were bored, such monotony.
They needed a change in ontogeny.
  To stay young at heart,
  They finally got smart,
And switched their life-style to neoteny.

NEOTENY

Neoteny, by definition,
Is a biotic condition,
  Wherein adult creatures
  Keep juvenile features,
Whence it becomes a tradition.

ONTOGENY

This is developmental biology,
To put it as simply as can be.
  From zygote to adult,
  Is what it's all about.
Or from acorn to an oak tree!

MORE ON ONTOGENY

"Ontogeny recapitulates phylogeny":
Used to be the popular phraseology.
  But then it was found
  That exceptions abound;
So, is it routine or mostly anomaly?

THE FATHER OF TAXONOMY

The Linnaean taxonomic hierarchy,
For its day was almost like anarchy.
  But for two-hundred years,
  It's conquered all fears,
It's not just a bunch of malarkey!

CAROLUS LINNAEUS

Kingdom, Phylum, Class, Order,
Family; but wait, it gets harder:
  Species, Subspecies, Genus,
  Who was this genius
Who tried to put God's house in order?

NOMENCLATURE

A species' nomenclature,
Can be difficult, to be sure.
  But its Latin name
  Can save the game...
By stating to whom we refer.

SYSTEMATICS

The practice of systematics,
You'd think would be automatic.
  But the cladist-phenetic war
  Was rampageous to its core,
And left the field highly chaotic!

ALTRUISM

Sacrifice and avoid egoism
Is the altruist catechism.
  Still, the final aim
  Is personal gain--
An ironic but all-too-truism!

EGOISM

With egos as big as the sea,
Some people act conceitedly.
  They're so horribly vain,
  It's almost insane:
They admire themselves more than me!

DETERMINISM
Some people think life is planned
By someone in total command:
   Donald J. Trump?
   No, he's a chump;
They're thinking of something more grand!

A SKEPTIC'S SKEPTICISM
Every scientist is a skeptic;
Against Faith it's an antiseptic.
   And if you believe this,
   Then I must insist...
You're not being scientific!

RELIGIOSITY
Here's the nitty-gritty
About religiosity:
   Faith and belief
   Offer relief...
Against much of life's adversity!

SPIRITUALITY
Scientists too can be spiritual,
Going through many a ritual;
   But is this just a mask
   We're compelled to ask?
Is reality real or just virtual?

ATHEISM AND AGNOSTICISM
Deep faith and religiosity,
Or data-driven curiosity?
   Ardent theism versus atheism,
   Are opposite sides of a schism
In-between which falls agnosticity!

PHILOSOPHIES ON LIFE
For some ambivalent folks,
Life is just a bunch of jokes.
   Every thought in our brain,
   Every joy and every pain,
Is all part of some grand hoax!

AMBIVOLENCE

Does ambivalence
Make much sense?
    I don't know or much care
    Anytime or anywhere
So, I'll just sit on the fence!

HUMAN LIFESPAN

The total temporal breadth
From birth until our death,
    Is called "lifespan"
    So, we must plan
Our first and every breath!

ON HER 100TH BIRTHDAY

Did she have a simple solution?
Some advice or wise conclusion?
    "How can we reach your age?"
    Her answer was truly sage:
"Be the tail of the distribution!"

DEATH

This event has wide renown.
As life's ill-fitting crown.
    Why death evolved
    Is a mystery unsolved,
But the message is clear: "Slow down!"

Genetics

G, NOW I C

I. O. what I. M. today
To C, T, G, and A.
  It's basically true
  I. M. me, U. R. U.,
Because of our D N A.

KIN SELECTION

The concept is highly infective,
But the phrase is rather deceptive--
  "Kin selection":
  The thought brings a grin.
You can pick your nose, not a relative!

LOVING FEELINGS

With emotions, genetics has toyed
Via how chromosomes are deployed.
  Meiosis created
  Our need to be mated,
By making our gametes haploid!

CULLULAR SEX

Making an ovum zygotic
Is a cellular process erotic.
  An egg weds a sperm
  In a union that's firm,
After which all becomes just mitotic!

© Springer International Publishing AG 2017
J.C. Avise, *From Aardvarks to Zooxanthellae*,
https://doi.org/10.1007/978-3-319-71625-1_15

AMAZING DEVELOPMENT

From the tips of our cute little noseies,
To the ends of our sweet tiny toesies,
    All cells are alike
    In base genotype.
They arose from a lot of mitoses.

TRANSPOSABLE ELEMENTS

These genes only do what is best
For their own selfish interest.
    With nary a care,
    They transpose here and there,
Unconcerned about making a mess.

THE WHY CHROMOSOME

Viewed through a geneticist's eye,
The topic of sex is quite dry.
    What most intrigues them
    Is not who or when;
They're focused on issues of Y.

DOS EQUIS

Another weird view about sex is:
How geneticists view their ex's.
    After his divorce,
    As a matter of course,
He ponders his loss of two X's.

TYPE-CAST

A genetics prof. full of hype,
Wooed a lovely coed phenotype.
    At the end of the term,
    Her rebuff was firm:
"Sorry, you're just not my geno-type."

FITNESS COMPONENTS

There are different kinds of ability.
Take the Mule, with his hybrid sterility.
    He's extremely tough,
    But his germs cells muff,
All pretentions of macho virility.

FITNESS FANATIC

To this tragedy, I was a witness.
The guy was genetically witless.
  He had muscles galore,
  But with girls didn't score,
Which made him a nut about fitness.

SEMINAL IDEA

To conceive of this operation,
Took a fertile imagination.
  In this form of husbandry,
  A syringe acts husbandly:
Artificial insemination.

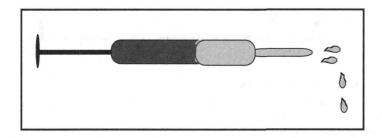

TEST TUBE BABIES

This form of fertilization
Has prompted some consternation.
  The gametic task
  Can take place in a flask,
While the parents are off on vacation!

ANGLES ON INHERITANCE

Horizontal genetic transmissions
Can occur via viral additions.
  But genes normally,
  Pass vertically,
When we take horizontal positions!

YOU CAN'T TWIN

For creating very close kin,
True cloning may seem like a sin.
  The mere thought of a clone
  Makes most of us moan;
But not an identical twin!

LET'S NOT GET RAILROADED

To train for this daring career
Demands a fast track and high gear,
    But we must ensure
    Loco-motives don't lure
The genetic engineer!

INBREEDING IS OUT

Societies must be repressive
Of marriage habits regressive.
    There's incest taboo
    'Cause it just doesn't do...
To wed deleterious recessives.

POPULATION GENETICS

This field can lead to obsession
With details of data compression.
    Do statistics aid
    The genetics trade?
Or lead to a lot of regression?

THE GENOMIC ERA

We've entered the age of genomics:
Transcriptomics, metabolomics,
    Proteomics, and other such names
    Yes, we're playing word games...
Much like a bunch of comics!

ACADEMIC OXYMORON

There must be some happy median,
Between the fields we're working in.
  So, every day,
  We combine work and play.
Like a serious comedian!

Anthropology: The Human Saga

THE HUMAN STORY

This will be the saga of man,
Told as best as I can.
My only intent
Is to make a small dent
In what you might not understand.

ANCESTRY OR MIMICRY?

Humans look quite like chimps, no debate.
Evolutionists claim we relate.
But Creationists say
God just made us that way;
But either way: we're a great ape!

EARLY PRIMATE EVOLUTION

The beginning of the tale,
Starts with the loss of the tail.
From Monkey to Ape,
That's all it would take.
For Primates to really set sail.

TALE OF TAILS

Monkey's tails are prehensile.
So, climbing trees was sensible.
But for Apes on the ground,
Tails hanging around
Would be silly and incomprehensible.

© Springer International Publishing AG 2017
J.C. Avise, *From Aardvarks to Zooxanthellae*,
https://doi.org/10.1007/978-3-319-71625-1_16

BIPEDALISM

When Apes came down from the trees,
They couldn't stand upright with ease.
  They weren't yet a biped,
  And therefore instead...
Kept skinning up their knees!

KNUCKLE-WALKING

Apes slowly mastered the notion:
Of this new form of locomotion.
  They best moved around,
  With hands on the ground.
Knuckle-walking was no loco-notion.

THE HUMAN-CHIMP SEPARATION

About six million years B.C.
From a split in our phylogeny.
  Most scientists propose
  That two lineages arose:
Proto-Human and pre-Chimpanzee.

THE BONOBO OR PYGMY CHIMP

There are two extant species of Chimp:
One big, the other a shrimp.
  The Bonobo,
  As you may know...
Is another name for the whimp.

AUSTRALOPITHECINES

Near the end of the Tertiary,
Just before the Quaternary.
  Australopithecus
  Was a cute little cuss:
A fossil most extraordinary!

THE NEXT STAGE

Let's put into proper context
What happened to happen next.
  About three million years ago,
  A genus labelled *Homo*...
Appeared in the evolutionary text.

MORE ON *HOMO*

Several different species of *Homo*
Through time, tended to come and go.
   Now there's only one left;
   So, the Earth is bereft.
Homo sapiens runs the whole show.

PLEISTOCENE

Dire Wolves and Sabre Cats mean,
Were just part of nature's team:
   Our contemporaries
   On ancient prairies…
Composing the Pleisto-scene.

HOMO SAPIENS

Turning the evolutionary pages,
Down throughout the ages.
   Two-hundred millennia ago,
   People like those we know.
Finally arrived on the stages.

NEANDERTHALS AND MODERN MAN

What about the Neanderthal?
The story may enthrall.
   Though different in size,
   We hybridized.
What do you think of that, y'all?

NEANDERTHAL LEGACY

It's stranger than it seems,
Even in our wildest dreams:
   From Neanderthal
   We've each and all…
Received four percent of our genes!

SKIN-DEEP DIFFERENCES

Exciting results just came in:
All races of man are close kin.
   Our family tree
   Proves genetically…
We're much alike under our skin.

MATERNAL ANCESTRY
A branched mitochondrial tree
Gives our matriline pedigree.
    At the dawn of time,
    Eve managed just fine
Acting great, great, ... grandmotherly.

THE Y CHROMOSOME
Quite analogously,
The Y's long history...
    Traces way back to Adam;
    And that's why we give a damn...
About patrilinearity!

OUR RELATIVELY RECENT PAST
Generations have come and gone:
About 20,000 strong!
    It's a long story...
    Filled with sadness and glory.
But I'll try not to make it too long.

WORRIED LOOKS
Our near ancestors walked a tough road.
On their minds was a heavy load.
    Their thick cranium
    Made them feel rather dumb;
So, their heavy brow ridges got furrowed.

MAMMOTH LIE
The Neanderthal fell for the con.
Now the Mammoth he'd cornered was gone.
    He believed the oaf,
    Who swore under oath:
"I'm not your prey-- I'm a Mastadon!"

HIGH STEAKS
After man had harnessed fire,
For Mammoths, things got dire.
    The sheer height of the beast
    Meant that each cooked feast.
Came from steaks that couldn't be higher!

UNWELCOME HOUSEGUEST

Early man could retreat without care,
To his cavern: a warm, cozy lair.
  Why humans left home
  Is not fully known,
But a prime suspect is the Cave Bear!

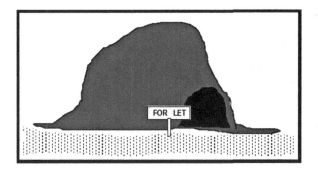

STONE AGE CULTURE

His stick was a garden mulcher,
And his stone could chase off a Vulture.
  Such inventions, though crude,
  Made it forever rude...
To suggest that man ain't got no culture!

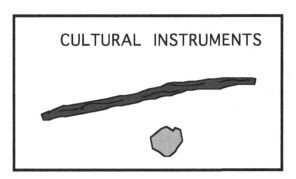

B*p*U*l*S*e*I*a*N*s*E*u*S*r*S*e

Human culture allowed for leisure,
But added also some pressure.
  As the title attests,
  It creates a big mess,
When you try to mix business with pleasure.

BEARING UP
> With nary a single appliance,
> On good sense they placed their reliance.
>> Without toasters or light,
>> Or TVs at night,
> Our forebearers practiced forebearance.

BE COOL
> In each warm, interglacial stage,
> Fresh food would rot, causing rage.
>> With no chilly fridge
>> To avoid such spoilage,
> Humans longed for each coming Ice Age.

PALEO-DIETS
> Speaking dietarily,
> We ate very warily…
>> As a carnivore, herbivore,
>> Omnivore, and even more,
> Almost necessarily!

EARLY INVENTOR
> People slowly but surely were yearning;
> Their quest for knowledge was burning.
>> Could they harness a fire?
>> Or maybe invent a tire?
> Their mental wheels got turning.

GOING TO SCHOOL
> Evolution took a big turn,
> And really started to churn,
>> After simple instinct,
>> Went nearly extinct,
> And mankind learned how to learn!

MAKING FRIENDS
> About ten thousand years ago,
> Give or take a decade or so,
>> Man domesticated
>> The wolves he had hated,
> Thereby making a best friend from foe.

CAT DOMESTICATION

Cat's too were domesticated,
The wisdom of which is debated:
   Avid Cat-lovers,
   Amid Dog-lovers:
War was bound to be generated!

MORE ON DOMESTIC CATS

The domesticated Cats,
Helped us kill lots of Rats.
   Given Cats' nine lives,
   After they arrived,
They put up impressive stats!

MAKING MONEY: A CAPITAL IDEA

Early businesses succeeded
When this simple motto was heeded:
   To make more bread
   Just use your head.
Lots of capital dough is kneaded."

DOUBLE ENTENDRES: FUR COATS AND SUNSCREEN

With a sense of humor and pride,
The chief announced to his tribe:
   "To conquer frostbite
   And avoid UV light,
Save your skin by tanning your hide!

ADVANCED LANGUAGE SKILLS

At language we were a whiz
Coining "It is what it is."
   "But it's not what it's not,
   Unless it is not
In which case it ain't what it is."

HUMAN BEINGS SOW AND TILL

Hunting resides in our genes,
But farming does not, so it seems.
  The rise of agriculture
  Fully changed our culture
And resulted in new human beans.

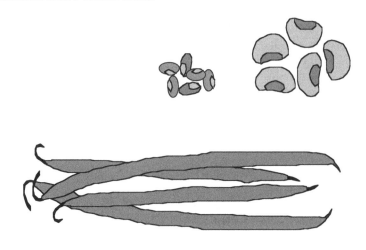

PLOWING AHEAD

In their first agronomics mode,
People labored hard, with backs bowed.
  They struggled until,
  They learned how to till.
'Twas a harrowing episode.

OPENING CLOTHING

An appealing concept quite deep,
Produced this insightful leap.
  Fig leaves were forgotten
  When man planted cotton
And learned: "You sew what you reap."

ECSTASY

Humans always sought X to C,
But instead, they found A to Z.
  When they realized
  They'd alphabetized,
They soon gained literacy!

ADVANCED LETTER TEXTING

The Cro-magnon said, "I.C.U."
Thus exhausting his words and I.Q.
From such base A.B.C.s,
To advanced Ph.D.s,
Human language has grown P.D.Q.

ADULT BEHAVIOR

Young humans are playful and boisterous.
Quite naturally girlish and boysterous.
When grown, it is found,
They still play around.
The actions just get more adulterous.

POLITICAL INCORRECTNESS

Though we're placed in the genus *Homo*,
Nowadays it's really a no-no…
To castigate,
Or denigrate,
By calling someone a "homo".

GAY PRIDE

Here's the shocking reality
About human sexuality:
Whether straight, bi-, or gay,
We can honestly say…
We love our *Homo*-sexuality.

CAPTIVATING KNOWLEDGE

Early women mastered the key
To attentive monogamy.
Though lacking college
Their carnal knowledge
Gained them grateful fidelity.

BARELY MAKING IT

We humans are quite splendiferous.
Among the Great Apes, least piliferous.
Our bare phenotype
Is a turn-on at night,
And explains why we've been so proliferous.

PRIMATES

Large males are clearly one fate
That female choice tends to create.
　　Being big and robust,
　　Enhances her lust,
As she searches hard for a prime-mate.

DARWINIAN SELECTION

Today, being slender is stressed,
But formerly, fatness was blessed.
　　Men and women searched hard
　　For mates made of lard.
Why?: "Survival of the fattest!"

COUNTRY CLUB

Early daters had no main hub;
No convening place, like a pub.
　　No lounges or bars;
　　Just love under stars,
When women went down to the club.

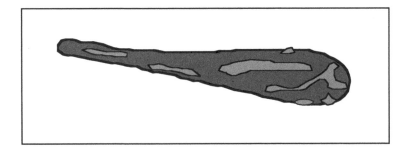

JEALOUSY

A Cro-Magnon husband dealt
A message that raised a welt.
　　When his wife fooled around
　　With a lover in town,
He gave her a chastity belt.

SOREHEAD

Her tone wasn't vaguely opaque.
She'd taken all she could take.
　　When clubbed on her head,
　　The phrase was first said:
"Not tonight, I've got a headache."

WOMEN'S LIBERATION

Modern women have taken a stance,
Saying "Mister, just give us a chance."
 But some men still wish you
 Would just skirt the issue,
Instead of wearing the pants.

RELIGION

Our world always has quite a few…
Who believe all they hear in a pew.
 Strict religious preachers…
 Are our species' teachers;
(Just between me and Ewe!).

MOVING MOVIES

We evolved in an Old World savannah;
East of Eden was somewhere near Kenya.
 From the place of our birth,
 Mankind moved 'round the earth;
In an Exodus: Out of Africa!

THE BERING STRAITS LAND BRIDGE

Through an Asian-Alaskan gate,
We came to the New World late.
 An Eskimo group
 Got lost in fog soup;
But soon got its bearings straight!

BRAVES' NEW WORLD

One thing I don't understand:
If Columbus discovered our land,
 Then why on the shore
 Could he not just ignore…
The welcoming Indian band?

BIG FOOTSTEPS

Paul Bunyan left thousands of lakes
From footprints across northern states.
 Why he never went south…
 Is debated about;
Perhaps his bunions got aches!

ONE LONG JOURNEY

It's been thousands of generations,
Countless cults, religions, and nations…
    Since humanity's birth
    On planet Earth.
It's taken a lot of patience!

THUMBTHING TO THINK ABOUT

Mankind a great distance has come:
To intelligent from rather dumb.
    What enabled this rise?
    Well, our cranial size;
Plus, of course, our opposable thumb!

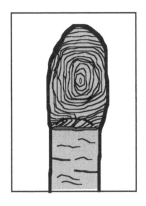

Appendix: Famous Animal Phrases

OUTLINE

This last section is a catch-all,
For creatures great and small…
    Who have been a subject
    (How could they object?)
For cute quotes by some know-it-all!

WOLF AND SHEEP

At first, he thought it was nothing,
But his mood changed to loathing,
    When a rancher saw and heard
    Rapidly approaching his herd,
A deadly *"Wolf in Sheep's clothing!"*

OLD DOGS AND NEW STICKS

He still loves to fetch my sticks,
And give my face lots of licks,
    But for a year or two,
    He's learned nothing new.
"You can't teach an old dog new tricks!"

FLY AND DOCTOR

Here's the note my dermatologist sent:
"I must cancel your next appointment."
    Then this wise-guy
    Explained just why:
"There's a Fly in the ointment!"

© Springer International Publishing AG 2017   147
J.C. Avise, *From Aardvarks to Zooxanthellae*,
https://doi.org/10.1007/978-3-319-71625-1

FISH AND AMPHIBIAN

On land, he was a new squatter;
No more use for a swim bladder.
  Earth's first Amphibian
  Was a piscine has-been,
Much "*like a Fish out of water!*"

SEABIRDS AND SEAMEN

The Captain fell to the deck,
Causing his vessel to wreck.
  There was nothing he could do
  About where the Seabird flew:
He had "*an Albatross around his neck!*"

LAMBS AND CLAMS

The recently-born Lamb
Was frolicking like a ham,
  Near his watchful Ewe,
  So, both of the two...
Were "*happy as a Clam!*"

DEER AND BUGS

A Fawn is a baby Deer
Born earlier in the year.
  For us and his Doe,
  It's almost as though...
"*He's as cute as a Bug's ear!*"

WASPS AND MAD HATTERS

It was surely no laughing matter,
When the Wasps began to scatter
  Away from their nest,
  Heading east and west,
And each one "*mad as a hatter!*"

GROUNDHOGS AND BUGS

When hibernating, this furry lug
Remains in a burrow she's dug;
  Each torpid Groundhog
  Will "sleep like a log":
"*As snug as a Bug in a rug!*"

HARES AND BEARS

When roused from its sleepy lair,
Any species of hibernating bear,
 Would act like another animal,
 Specifically, a small mammal:
They'd be *"as mad as a March Hare!"*

CROWS AND PLANES

When traversing friendly skies,
Birds, like airliners, find it wise
 To journey here and about
 By a least circuitous route:
They go *"as straight as the Crow flies!"*

HUSBANDS AND WIVES

He incurred her seething wrath
Before breathing his last breath;
 His mammalian mate
 Had sealed his ill fate,
By *"Badgering him to death!"*

DOGS AND SQUIRRELS

It's very hilarious to see
Any Dog acting frantically…
 Chasing a ground squirrel,
 And getting all in a whirl:
Clearly *"barking up the wrong tree!"*

BATS AND ARBORISTS

When the arborist cut down a tree,
The ignoramus simply didn't see…
 Any need to stand clear;
 Was he just lacking fear?
Or perhaps had *"Bats in the belfry!"*

BEES AND APIARISTS

This is an obvious sonnet,
So, I'll get straight on it:
 The girl apiarist
 Got very pissed
When she got *"Bees in her bonnet!"*

BIRDS AND MORE BIRDS
It doesn't matter whether,
It's stormy or sunny weather;
   Black sky or blue,
   It's forever true:
"Birds of a feather flock together!"

SHEEP AND FAMILIES
In everyone's history or pedigree,
Someone was odd, to some degree.
   Whether black or white,
   Something wasn't quite right:
'Bout the *"black Sheep of the family!"*

CATS AND GUYS
What's another word for dung?
Or for a guy who's well hung?
   Are you circumspect,
   (Which I fully respect),
Or *"has the Cat got your tongue?"*

BULL AND BULLFIGHTER
As a Bull, are you a stud?
Or merely more of a dud?
   Will you use your great might…
   In tomorrow's Bullfight?
Or just lay there *"chewing your cud?"*

CHICKENS AND FARMERS
Rather than be loosed,
Or even gently noosed,
   On our huge family farm,
   We know there's less harm,
When our *"Chickens come home to roost!"*

ELEPHANTS AND ROOMS
Are we really safe to assume
There's no impending doom?
   But I must yell:
   What the hell!!
"There's an Elephant in the room!"

NEWTS AND WITCHES

What's in that witches' brew?
You only wish you knew!:
  "Eye of Newt, toe of Frog,
  Wool of Bat, tongue of Dog";
It's enough to make you puke!

HORSES AND JOCKEYS

It was horribly stupid, of course,
Sheer frustration was its source;
  The jockey had made his point,
  Even though there was <u>no</u> point
To *"flogging a dead Horse"*.

FLIES AND SPIES

The spy answered the call,
And he had no lack of gall.
  Though there wasn't much room,
  He sneaked into the boardroom
"Just like a Fly on the wall!"

HORSES AND DICTATORS

Every dictator, of course,
Acts nasty and very coarse;
  Everyone needs to say,
  Don't behave that way!
Just *"Get off your high Horse!"*

GOATS AND LAUREATES

The winner started to gloat,
He was a personage of note.
  The Noble Prize
  Shown in his eyes;
He was *"giddy as a Goat!"*

HOGS AND RACERS

Many miles he started to log,
As he lengthened out his jog;
  Could he go any faster?
  He was trying to master...
The art of *"going in whole Hog!"*

DOGS AND PET STORES
Some pet-store workdays are slogs,
You feel like you're mired in bogs;
   You question yourself
   With a query heartfelt:
Am I really *"going to the Dogs?"*

CATS AND MAD-HATTERS
He's rather obnoxious and fat;
He might wear a mad-hatter's hat;
   His face-wide smile
   Masks a lot of guile;
He *"grins like a Cheshire Cat!"*

HORSES AND PROFESSORS
Teaching too many tough courses,
Can drain all your mental resources
   So, each Professor
   Has got to be sure
To sometimes *"hold your Horses!"*

HORSES AND CARTS
You must take this to heart:
Be correct, right from the start.
   If you do it wrong,
   You'll get the gong!
"Put the horse before the cart!"

FROGS AND CATS
It's great weather for a Frog,
Who loves to sit in the sog.
   You can bet
   It's very wet…
When *"raining Cats and Dogs!"*

HERRINGS AND SHOPPERS
He thought he had gotten his bearing,
He even stopped cussing and swearing!
   He had sought a fish market,
   But was still way off target.
What he spotted was just a *"red Herring!"*

COWS AND COOKS

Stop that cooking right now!
To authority you must bow.
  Don't ever eat
  Un-kosher meat
Or that from a "*sacred Cow!*"

BEES AND SPELLING BEES

How do you spell "B"?
It's as simple as can be;
  But one letter or two,
  May still be too few…
To win a "*spelling-bee!*"

PIGEONS AND THE ENGLISH

These "birds" are legion,
In every English region;
  On almost every street,
  Half the people you meet…
Might be a "*stool Pigeon!*"

SWANS AND SONGS

Unless I'm totally wrong,
Although his neck is long,
  He has no vocal cord,
  And, thus, gets ignored,
During his final "*Swan song!*"

BEES AND KNEES

The phrase was spoken with ease,
But tell me, if you would, please:
  What does it mean?
  To know, I'm keen:
What in the world are "*the Bee's knees*"?

MICE AND MEN

Every now and again,
I wonder how and when,
  Plus, where and why
  Are thought to lie…
"*the best schemes of Mice and men*".

BIRDS AND BEES

I thought it would be a breeze,
But when she asked, "Dad please,
   Tell me how to conceive"
   I just couldn't believe...
I said query "*the Birds and the Bees!*"

DOGS AND TAILS

I once had a hunting Dog
Who slept all day like a log,
   When he did wag his tail,
   It almost seemed to foretell
That "*the tail was wagging the dog!*"

COWS AND RANCHERS

No matter how far they roam,
Or how many fields they comb,
   At the end of the day,
   They must find their way.
So, we wait "*till the Cows come home!*"

GEESE AND HUNTERS

It's always at a fast pace,
Whoever wins this race.
   For this strange endeavor,
   You must be quite clever,
Or it ends as a "*wild Goose chase!*"

COWS AND MILKERS

She looked as if to say "wow"!
Her milker didn't know how;
   So, about his task
   He had to ask:
"*How now, brown Cow?*"

GUPPIES AND PUPPIES

A house full of yuppies
Had many quiet Guppies,
   But their five baby Dogs
   Drove all of them agogs;
So, they often said: "*hush, Puppies!*"

PIGS AND SAILORS

Do you think that bye and bye,
He'll repay his loan? "Aye aye"...
   Said the sailor with a smile
   'Cause he knew all the while...
That would happen *"in a Pig's eye!"*

GROUSE AND LOUSE

He'd acted like a Louse;
Grumpy as a Grouse.
   Tonight, his sleep
   Would not be deep.
He was *"in the Doghouse"*.

ORCAS AND OTTERS

The Orcas caught her:
That stupid Sea Otter.
   She'd paddled offshore
   Only to encounter gore.
"Like a Lamb going to the slaughter!"

DUCKS AND POLITICS

He finally had run out of luck
Though gimpy, he still had pluck
   At the waterfowl convention,
   He wasn't the flock's selection.
So, for now, he's just *"a lame Duck"*.

CATS AND BAGS

At first, it seemed just a cag,
With surely no warning flag.
   But then, his friends said:
   "You might end up dead".
So, they *"let the Cat out of the bag!"*

CHICKENS AND EXECUTIONERS

His initial reaction was to scoff,
When he heard it from his Prof.
   How could a bird run around
   With its head on the ground?
"like a Chicken with his head cut off."

MOTHS AND CANDLES

She was his first dame,
In nature's mating game;
  Sweet and demure,
  He was drawn to her…
"Like a Moth to a flame!"

OWLS AND NIGHT-OWLS

In weather fair or foul,
They never wear a scowl;
  Though some are diurnal,
  Most are mostly nocturnal;
So, it's proper to say *"night-Owl!"*

TIGERS AND ORAGAMI

This cat hunts with famed vigor;
Chasing prey, it has a high gear.
  But as folded Oregami,
  You get what you see:
Only a *"paper Tiger!"*

HORSES AND GIRLS

He helped the little gal up;
And the girl rode off at a gallop.
  Though not a full horse,
  She had, of course,
At least started to *"Pony-up!"*

PARTRIDGES AND TREES

This phrase brings glee,
Evoking sentimentality,
  At Christmas time,
  When we chime:
"A Partridge in a pear tree."

PIGS AND MOONS

His sense of humor was wry,
Quite sensible, yet rather dry.
  Will we colonize the moon?
  "Yes, I'm sure pretty soon":
Whenever *"Pigs might fly!"*

THE END

That's biology, in all its great glory;
From marvelous and sublime to gory;
    Its telling has been quite a feat,
    And although it's not complete,
This will mark the end of my story!

.

Printed in the United States
By Bookmasters